W9-ARQ-900

Contents at a Glance

Table of Contents

SPSS®
Statistics
FOR
DUMMIES®
A Wiley Brand

3rd Edition

by Keith McCormick and Jesus Salcedo
with Aaron Poh

SPSS® Statistics For Dummies®, 3rd Edition

Published by: **John Wiley & Sons, Inc.,** 111 River Street, Hoboken, NJ 07030-5774,
www.wiley.com

Copyright © 2015 by John Wiley & Sons, Inc., Hoboken, New Jersey

Published simultaneously in Canada

For general information on our other products and services, please contact our Customer Care Department within the U.S. at 877-762-2974, outside the U.S. at 317-572-3993, or fax 317-572-4002. For technical support, please visit www.wiley.com/techsupport.

Wiley publishes in a variety of print and electronic formats and by print-on-demand. Some material included with standard print versions of this book may not be included in e-books or in print-on-demand. If this book refers to media such as a CD or DVD that is not included in the version you purchased, you may download this material at http://booksupport.wiley.com. For more information about Wiley products, visit www.wiley.com.

Library of Congress Control Number: 2015940308

ISBN 978-1-118-98901-2 (pbk); ISBN 978-1-118-98902-9 (ebk); ISBN 978-1-118-98903-6 (ebk)

Manufactured in the United States of America

10 9 8 7 6 5

Introduction

....

Good news! You don't have to know diddlysquat about the math behind statistics to be able to come up with well-calculated conclusions and display them in fancy graphs. We won't be doing any calculations by hand. All you need is the IBM SPSS Statistics software and a bunch of numbers. This book shows you how to type the numbers, click options in the menus, and produce brilliant statistics. And interpret them properly, too! It really is as simple as that.

About This Book

This is fundamentally a reference book. Parts of the book are written as stand-alone tutorials to make it easy for you to get into whatever you're after. Once you're up and running with SPSS, you can skip around and read just the sections you need. You really don't want to read straight through the entire book. That way leads to boredom. We know — we went straight through everything to write the book, and believe us, you don't want to do that.

This book is not about math. It's about statistics. You don't derive anything. You don't do any math by hand or look up numbers in statistical tables. You won't find one explanation of how calculations are performed under the hood. This book is about the things you can do to command SPSS to calculate statistics for you. The inside truth is that you can be as dumb as a post about statistical calculation techniques and still use SPSS to produce some nifty stats!

However, if you decide to study the techniques of statistical calculation, you'll be able to understand what SPSS does to produce numbers. Your main advantage in understanding the process to that degree of detail is that you'll be able to choose a calculation method that more closely models the reality you're trying to analyze — if you're interested in reality, of course.

Throughout the book you find examples that use data stored in files. These files are freely available to you. Most of the files are installed with IBM SPSS Statistics in the SPSS installation directory, which, by default, is \Program Files\SPSS (unless you chose another location during installation). A few files were designed for this book and are available on the book's companion website (see "Beyond the Book" for more information). In every case, the files were especially designed to demonstrate some specific capability of SPSS.

Within this book, you may note that some web addresses break across two lines of text. If you're reading this book in print and want to visit one of these web pages, simply key in the web address exactly as it's noted in the text, pretending as though the line break doesn't exist. If you're reading this as an e-book, you've got it easy — just click the web address to be taken directly to the web page.

Finally, a technical note: The official name of the product is IBM SPSS Statistics. Throughout this book, we refer to it simply as SPSS. Outside of this book, that shortcut can be risky because there are other related products also called SPSS — notably, IBM SPSS Modeler, which, though powerful and part of the same brand, is not the subject of this book.

Foolish Assumptions

This book is for anyone new to SPSS. No prior knowledge of statistics or mathematics is needed or even expected. In specific terms, we made a few assumptions about you, the reader of this book:

- ✔ You may be a student who isn't majoring in mathematics but has been instructed to use SPSS by one of your professors.
- ✔ You may be an office worker who has been told to use SPSS to analyze some data.

For most people who generate statistics, the complexity of using the software becomes an obstacle. Our purpose in writing this book is to show you how to move that obstacle out of the way with minimum effort.

Icons Used in This Book

Throughout this book, we use icons in the margins to grab your attention. Here's what those icons mean:

You should keep this information in mind. It's important to what you're doing.

This icon highlights unnecessarily geeky information, but we had to include it to complete the thought. You can skip anything marked with this icon unless the text makes you curious.

This icon highlights a point that can save you time and effort.

Anything marked with this icon offers information about something that can sneak up and bite you.

Beyond the Book

In addition to the material in the print or e-book you're reading right now, this product also comes with some access-anywhere goodies on the web. Check out the free Cheat Sheet at www.dummies.com/cheatsheet/spss for information on variable levels of measurement, commonly used procedures within the Analyze menu, and possible conclusions that you can reach after conducting a statistical test. We also provide free articles on the web at www.dummies.com/extras/spss, on topics such as automatic recode, creating your own table look, and more. Finally, you can download data files that don't come with SPSS at www.dummies.com/go/spss.

Where to Go from Here

We recommend starting out by reading Chapter 1, so you understand what SPSS is. (We tried to leave out the boring parts.) If you haven't already installed SPSS, check out Chapter 2. Read the stuff in Chapter 4 about defining variables and entering data — it all makes sense once you get the hang of it, but the process seems kind of screwy until you see how it works. And from there, use the Table of Contents and Index to find the things you want to do!

If you have a question about the data, or if you want to contact us about some other question you may have, you can reach Keith McCormick at keithmc123@gmail.com.

Part I
Getting Started with SPSS

In this part . . .

- ✔ Find out what SPSS is all about.
- ✔ Install SPSS properly and customizes the settings.
- ✔ Get a feel for SPSS with a very simple first example session.

Chapter 1

Introducing SPSS

A statistic is a number. A *raw* statistic is a measurement of some sort. It's fundamentally a count of something — occurrences, speed, amount, or whatever. A statistic is calculated using a sample. In a sense, a sample is the keyhole you have to peer through to the population, which is what you're trying to understand. The value at the population level — the average height of an American male, for instance — is called a *parameter*. Unless you've got all the data there is, and you've collected a census of the population, you have to make do with the data in your sample. The job of SPSS is to calculate. Your job is to provide a good sample.

In this chapter, we discuss the importance of having accurate, reliable data, and some of the implications when this is not the case. We also talk about how best to organize your data in SPSS and the different kinds of files that SPSS creates. We take a trip down memory lane and discuss the origins of SPSS, as well as what can be done in the program and different ways of communicating with the software. Finally, we spend some time discussing different ways in which you can get help when navigating SPSS.

Garbage In, Garbage Out: Recognizing the Importance of Good Data

SPSS doesn't warn you when there is something wrong with your sample. Its job is to work on the data you give it. If what you give SPSS is incomplete or biased, or if there is data that doesn't belong in there, the resulting calculations won't reflect the population very well. Not much in the SPSS output will signal to anyone that there is a problem. So, if you're not careful, you can conclude just about anything from your data and your calculations.

Consider the data in Table 1-1. What if you calculated the survival rate of *Titanic* passengers based on this small sample? What if you calculated what fraction of the passengers were in each class of service? You can easily see that you'd be in real trouble.

However, consider this: Would you be tempted to drop these cases from your analysis because their fare information appears to be missing? What if fare information were provided for all the other passengers? You might drop the cases in Table 1-1 but use everyone else. You'd be dropping only a handful of passengers out of hundreds, so that would be okay, right? The answer is no, it would not be okay. As it turns out, there is a good reason that each of these passengers didn't pay a fare (for example, Mr. Thomas Andrews, Jr., designed the ship), and if this was your data, your job would be to know that.

Sampling is a big topic, but here's the quick version:

- ✔ The data points in your sample should be drawn at random from the population.
- ✔ There should be enough data points.
- ✔ You should be able to justify the removal of any data points.

This book is not about the accuracy, correctness, or completeness of the input data. Your data is up to you. This book shows you how to take the numbers you already have, put them into SPSS, crunch them, and display the results in a way that makes sense. Gathering valid data and figuring out which crunch to use is up to you.

Your data is your most valuable possession, so be sure to back it up. Make sure you have multiple backups, with at least one stored offsite. The last thing you want is to lose your data.

Table 1-1		Sample of *Titanic* Passengers					
Survived or Died	**Class**	**Name**	**Sex**	**Age**	**Fare Paid**	**Cabin**	**Embarkation**
Died	1	Andrews, Mr. Thomas, Jr.	Male	39	0.00	A36	Southampton
Died	1	Parr, Mr. William Henry Marsh	Male		0.00		Southampton
Died	1	Fry, Mr. Richard	Male		0.00	B102	Southampton
Died	1	Harrison, Mr. William	Male	40	0.00	B94	Southampton
Died	1	Reuchlin, Mr. John George	Male	38	0.00		Southampton
Died	2	Parkes, Mr. Francis "Frank"	Male		0.00		Southampton
Died	2	Cunningham, Mr. Alfred Fleming	Male		0.00		Southampton
Died	2	Campbell, Mr. William	Male		0.00		Southampton
Died	2	Frost, Mr. Anthony Wood "Archie"	Male		0.00		Southampton
Died	2	Knight, Mr. Robert J.	Male		0.00		Southampton
Died	2	Watson, Mr. Ennis Hastings	Male		0.00		Southampton
Died	3	Leonard, Mr. Lionel	Male	36	0.00		Southampton
Died	3	Tornquist, Mr. William Henry	Male	25	0.00		Southampton
Died	3	Johnson, Mr. William Cahoone, Jr.	Male	19	0.00		Southampton
Died	3	Johnson, Mr. Alfred	Male	49	0.00		Southampton

The origin of SPSS

SPSS is probably older than you are. In 2018, it will turn 50. That makes it older than Windows and older than the first Apple computer, so in the early days SPSS was run on mainframe computers using punch cards.

At Stanford University in the late 1960s, Norman H. Nie, C. Hadlai (Tex) Hull, and Dale H. Bent developed the original software system named Statistical Package for the Social Sciences (SPSS). They needed to analyze a large volume of social science data, so they wrote software to do it. The software package caught on with other folks at universities, and, consistent with the open-source tradition of the day, the software spread through universities across the country.

The three men produced a manual in the 1970s, and the software's popularity took off. A version of SPSS existed for each of the different kinds of mainframe computers in existence at the time. Its popularity spread from universities into the public sector, and it began to leak into the private sector as well.

In the 1980s, a version of the software was moved to the personal computer. In 2008, the name was briefly changed to Predictive Analytics Software (PASW). In 2009, SPSS, Inc., was acquired by IBM, and the name of the product was returned to the more familiar SPSS. The official name of the software today is IBM SPSS Statistics.

SPSS is available in several forms — single user, multiuser, client-server, student version, and so on. The software also has a number of special-purpose add-ons. You can find out about them all at www-01.ibm.com/software/analytics/spss/products/statistics.

Talking to SPSS: Can You Hear Me Now?

More than one way exists for you to command SPSS to do your bidding. You can use any of four approaches to perform any of the SPSS functions, and we cover them all in this section. The method you should choose depends not only on which interface you prefer, but also (to an extent) on the task you want performed.

The graphical user interface

SPSS has a window interface. You can issue commands by using the mouse to make menu selections that cause dialog boxes to appear. This is a fill-in-the-blanks approach to statistical analysis that guides you through the process of making choices and selecting values. The advantage of the graphical user interface (GUI) approach is that, at each step, SPSS makes sure you enter everything necessary before you can proceed to the next step. This interface is preferred for those just starting out — and if you don't go into depth with SPSS, this may be the only interface you ever use.

Syntax

Syntax is the internal language used to command actions from SPSS. It's the command syntax of SPSS (hence, its name). Syntax is often referred to as the "command language." You can use the Syntax command language to enter instructions into SPSS and have it do anything it's capable of doing. In fact, when you select from menus and dialog boxes to command SPSS, you're actually generating Syntax commands internally that do your bidding. In other words, the GUI is nothing more than the front end of a Syntax command-writing utility.

Writing (and saving) command-language programs is a good way to create processes that you expect to repeat. You can even grab a copy of the Syntax commands generated from the menu and save them to be repeated later.

Python programs

Python is a general-purpose language that has a collection of SPSS modules written for it; you can use Python to write programs that work inside SPSS. You can also run Python with the Syntax language to command SPSS to perform statistical functions.

One advantage of Python is that it's a modern language, complete with the power and convenience that come with such languages, including the capability of constructing a more readable program. In addition, because Python is a general-purpose language, you can read and write data in other applications and files. Think of Python programs as a way of making Syntax more powerful.

Python scripts

What SPSS calls *scripts* are also written in Python, but they help you manipulate the GUI. They're a little more advanced and quite powerful. You use Python scripts to automatically highlight certain results in the SPSS output, for instance.

How SPSS Works

The developers of SPSS have made every effort to make the software easy to use. SPSS prevents you from making mistakes or even forgetting something. That's not to say it's impossible to do something wrong in SPSS, but the SPSS software works hard to keep you from running into the ditch. To foul things up, you almost have to work at figuring out a way of doing something wrong.

You always begin by defining a set of *variables;* then you enter data for the variables to create a number of *cases.* For example, if you're doing an analysis of automobiles, each car in your study would be a case. The variables that define the cases could be things such as the year of manufacture, horse-power, and cubic inches of displacement. Each car in the study is defined as a single case, and each case is defined as a set of values assigned to the collection of variables. Every case has a value for each variable. (Well, you *can* have a missing value, but that's a special situation described later.)

Each variable is a specific type. Types describe how the data is *stored* — for example, as letters (strings), as numbers, as dates, or as currency (see Chapter 4 for more information on data types). Each variable is defined as containing a certain kind of number, so you also have to define the variable's level of measurement. For example, a *scale* variable is a numeric measure-ment, such as weight or miles per gallon. A *categorical* variable contains values that define a category; for example, a variable named gender could be a categorical variable defined to contain only values 1 for female and 2 for male. Things that make sense for one type of variable don't necessarily make sense for another. For example, it makes sense to calculate the average miles per gallon, but not the average gender.

After your data is entered into SPSS — your cases are all defined by values stored in the variables — you can easily run an analysis. You've already finished the hard part. Running an analysis on the data is simple compared to entering the data. To run an analysis, you select the analysis you want to run from the menu, select the appropriate variables, and click OK. SPSS reads through all your cases, performs the analysis, and presents you with the output as tables or graphs. Of course, you have to know which analysis to chose. For that, too, we have you covered (see Part V).

You can instruct SPSS to draw graphs and charts directly from your data the same way you instruct it to do an analysis. You select the desired graph from the menu, assign variables to it, and click OK.

When you're preparing SPSS to run an analysis or draw a graph, the OK button is unavailable until you've made all the choices necessary to produce output. Not only does SPSS require that you select a sufficient number of vari-ables to produce output, but it also requires you to choose the right kinds of variables. If a categorical variable is required for a certain slot, SPSS won't allow you to choose any other kind of variable. Whether the output makes sense is up to you and your data, but SPSS makes sure that the choices you make can be used to produce some kind of result.

Numbers not words

SPSS works best with numbers. Whenever possible, try to have your SPSS data in the form of numbers. If you give SPSS names and descriptions, it'll seem like they're being processed by SPSS, but that's because each name has been assigned a number. (Sneaky.) That's why survey questions are written like this:

How do you feel about rhubarb? Select one answer:

A. I love it!

B. It's okay.

C. I can take it or leave it.

D. I don't care for it.

E. I hate it!

A number is assigned to each of the possible answers, and these numbers are fed through the statistical process. SPSS uses the numbers — not the words — so be careful about keeping all your words and numbers straight. We cover this subject in some detail in Chapter 4.

Remember: Keep accurate records describing your data, how you got the data, and what it means. SPSS can do all the calculations for you, but only you can decipher what it means. In *The Hitchhiker's Guide to the Galaxy,* a computer the size of a planet crunched on a problem for generations and finally came out with the answer, 42. But the people tending the machine had no idea what the answer meant because they didn't remember the question. They hadn't kept track of their input. You must keep careful track of your data or you may later discover, for example, that what you've interpreted to be a simple increase is actually an increase in your rate of decrease. Oops!

All output from SPSS goes to the same place — a dialog box named SPSS Statistics Viewer. This dialog box displays the results of whatever you've done. After you've produced output, if you perform some action that produces more output, the new output is displayed in the same dialog box. And almost anything you do produces output. Of course, you need to know how to interpret the output — SPSS will help you, and so does this book.

Making sense of all those SPSS files

Input data and statistics are stored in files — different kinds of files. Some files contain numbers and definitions of numbers. Some files contain graphics. Some files contain both. Data files are easy to spot because they end with the extension `.sav`. Output files end with the extension `.spv`. Command Syntax files, with the optional programming language commands, end with `.sps`.

The examples in this book require the use of files that contain data configured to demonstrate capabilities of SPSS. Most of the files are in the same directory you used to install SPSS (installing SPSS also installs a number of data files ready to be loaded into SPSS and used for analysis). A few of the files used in the examples can be downloaded from the book's companion website (`www.dummies.com/go/spss`).

Getting Help When You Need It

You're not alone. Some immediate help comes directly from the SPSS software package. More help can be found online. If you find yourself stumped, you can look for help in several places:

- ✔ **Topics:** Choosing Help➪Topics from the main window of the SPSS application is your gateway to immediate help. The help is somewhat terse, but it usually provides exactly the information you need. The information is in one large Help document, presented one page at a time. Choose Contents to select a heading from an extensive table of contents, choose Index to search for a heading by entering its name, or choose Search to enter a search string inside the body of the Help text.

  In the Help directory, the titles in all uppercase are descriptions of Syntax language commands.

- ✔ **Tutorial:** Choose Help➪Tutorial to open a dialog box with the outline of a tutorial that guides you through many parts of SPSS. You can start at the beginning and view each lesson in turn, or you can select your subject and view just that.

- ✔ **Case Studies:** Choose Help➪Case Studies to open a dialog box containing examples in a format similar to that of the Tutorial. You can select titles from the outline and view descriptions and examples of specific instances of using SPSS. You can also find descriptions of the different types of calculations. If some particular analysis type is eluding your comprehension, this is a good place to look.

- ✔ **Statistics Coach:** Choose Help➪Statistics Coach if you have a good idea of what you want to do but you need some specific information on how to go about doing it.

- ✔ **Command Syntax Reference:** Choose Help➪Command Syntax Reference to display more than 2,000 pages of references to the Syntax language in your PDF viewer. The regular Help topics, mentioned earlier, provide a brief overview of each topic, but this document is much more detailed.

- ✔ **Algorithms:** Choose Help➪Algorithms to get detailed information on how processes work internally. This is where you can dive far down into the internals. If you want to take a look at the math and how it's applied, this is where you should start.

Chapter 2

Installing SPSS

*T*his chapter is all about installing your software and setting the options that determine how it works. If the software you'll be using is already installed, you can skip the first part of this chapter and jump right to configuration a little further on.

The installation process guides you step by step and then does most of the work itself. The configuration settings all default to something reasonable, so we suggest leaving them alone for now. You can always come back later and make a change if you develop a gripe.

Getting SPSS onto Your Computer

Soap powder comes in boxes, paint comes in cans, corn dogs come on sticks, and SPSS comes on the Internet. SPSS used to come on a CD, but it's now available only as an Internet download. It's still the same software — the only real difference is where the files come from.

When you download the SPSS software off the Internet, find a place to put the files and all its contents on your hard drive. Don't throw out anything. Make sure to keep meticulous records of the website you downloaded from, which files you downloaded, and all numbers and identifiers you encounter. Trust us, you'll need them later.

The Mac and Linux versions of the software are similar in operation, but details of the installation procedure described here are specific to Windows.

What you need for running SPSS

You won't have to worry about the minimum requirements for the computer, unless yours is an antique. After all, who *doesn't* have at least 256MB of RAM and 300MB of free disk space?

SPSS comes in a variety of flavors. They're fundamentally alike, but some versions have more parts than others. You may have all, some, or none of the add-ons described in Chapter 22. In any case, you need an authorization code to enable whatever you do have. You'll need to authorize your base system, as well as any add-ons. You may have more than one authorization code — it depends on how your SPSS system is configured, which is determined by what parts are included with it.

More than one version of IBM SPSS Statistics 23 exists, for execution under different operating systems. IBM SPSS Statistics 23 for Windows can be run on a variety of Windows platforms, including Windows 8.1 and Windows 7 in either 32-bit or 64-bit. You can run IBM SPSS Statistics 23 for Mac on Macintosh 10.10 (Yosemite) or most recent versions of OS X in 64-bit. IBM SPSS Statistics 23 for Linux has been tested on Red Hat Enterprise Linux 6 and Debian 6.0, but it should also run on any sufficiently updated Linux system. Go to `www-01.ibm.com/software/analytics/spss/products/ statistics/requirements.html` to check the detailed system requirements for SPSS.

For the installation procedure to work, you must be logged into your Windows system with administrator privileges. You don't have to be logged in as an administrator, but whatever login you're using must have the privileges that the administrator has.

You should also be connected to the Internet. You can install SPSS without being connected, but it's a pain to do it that way. Make it easy on yourself and connect your computer to the Internet before you start. And keep it connected at least until you get SPSS installed.

In summary, before you begin the installation:

✔ You must have access to your authorization code or codes.

✔ You must have access to the serial number of your copy of SPSS.

✔ You may also need to have access to your customer number.

✔ You must be logged into your computer with administrator privileges.

✔ For convenience, you probably want to be connected to the Internet.

Cranking up the installer

The installation procedure is dead simple: You simply start the installation program and answer the questions. And the questions are easy.

If you have a previous version of the software installed, you may want to remove it before you install the new version. Some folks opt to have both versions running for a time to test old Syntax. This can get a little confusing, but it isn't uncommon.

To remove the previous version, use the Windows Control Panel and select the old version of SPSS. Then click the Remove button or Uninstall button to delete it.

You can start the installer after you've downloaded the SPSS software off the Internet. If you've downloaded your version of the software, you've gotten an executable program. All you need to do is run it. The first dialog box you see is shown in Figure 2-1. Click Next.

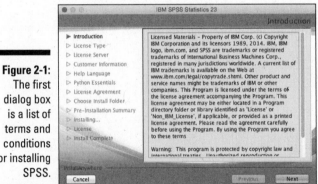

Figure 2-1:
The first dialog box is a list of terms and conditions for installing SPSS.

As you can see in Figure 2-2, you install SPSS according to the type of license you've purchased. The example described in this chapter is for an authorized single-user installation, but you can also install it under a site license or a network license.

The SPSS installation sequence

With the dialog box shown in Figure 2-2 on your screen, select the type of license you have and click Next. After you make your selection, you're

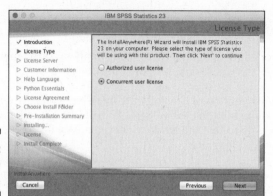

Figure 2-2:
Choose the
license type.

greeted by the dialog box shown in Figure 2-3. Enter the license manager
name or IP address and click Next to move forward.

Figure 2-3:
Enter your
license
manager
name or
server IP
address.

The next dialog box, shown in Figure 2-4, asks for your name and organization. (You can put anything you like in the Organization field, but keep it
clean, because it could pop up on the screen one afternoon while your mom
is watching.)

When you click Next, you're asked if you want to install the Essentials for
Python package, as shown in Figure 2-5. You can decide for yourself if you
want to use Python with SPSS. If you can't think of a reason you would need
it, select No and move on by clicking Next.

Essentials for Python does allow you additional options and access to more
techniques, so you may want to consider installing this.

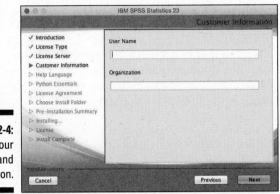

Figure 2-4:
Enter your
name and
organization.

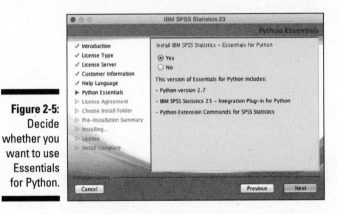

Figure 2-5:
Decide
whether you
want to use
Essentials
for Python.

After you make your selection, you're greeted by the license agreement,
as shown in Figure 2-6. Simply do what it says: Read the license, and if you
accept the terms, select the I Accept the Terms of the License Agreement
option and then click Next.

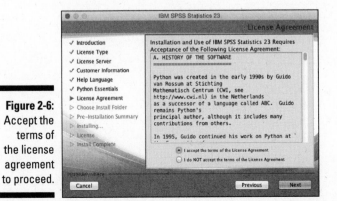

Figure 2-6:
Accept the
terms of
the license
agreement
to proceed.

The dialog box that appears at this point asks whether you really want to install SPSS. All you've done so far is answer some questions; nothing has been installed. This dialog box has a Previous button you can click to go back and change your answers. The Next button unleashes the installation software onto your computer. The dialog box also has a Cancel button if you chicken out, or if you enjoyed the process so much that you want to drop everything and do the entire thing all over again. If you actually want SPSS on your computer, click Next.

The next dialog box, shown in Figure 2-7, lists every file being installed, while a progress indicator moves across the screen. The filenames flicker by pretty fast; only Superman or Data from *Star Trek* could read them. Normal mortals see mostly a line of constantly flickering letters.

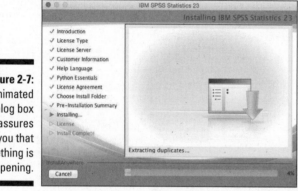

Figure 2-7: An animated dialog box reassures you that something is happening.

The progress indicator marches across the screen until it reaches the far right. At that point, the flickering of filenames will stop. For a time, nothing moves. Be patient. Just about the time you start to wonder whether something has gone wrong, you see the dialog box shown in Figure 2-8.

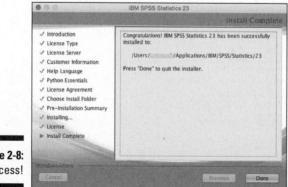

Figure 2-8: Success!

The last dialog box that you see is Install Complete. Click Done. The dialog box disappears, and your software is ready to run.

Late registration

If you installed SPSS but chose to register it later, or if you want to check the status of your registration, you can do that easily. Simply choose Start⇨All Programs⇨IBM SPSS Statistics⇨IBM SPSS Statistics 23 License Authorization Wizard, and the status of your license appears onscreen. Then you get the same sequence of registration dialog boxes described in the previous section.

The Internet being the Internet, your connection may get dropped right in the middle of the registration process. If that happens, just start over from the Start menu.

Starting SPSS

You now have SPSS installed on your computer. You'll find a listing for it with the other programs in your Start menu. Choose Start⇨All Programs⇨IBM SPSS Statistics. You then have two choices:

- ✔ IBM SPSS Statistics 23
- ✔ IBM SPSS Statistics 23 License Authorization Wizard

The first choice is the main program itself — and that will be the number-one selection on your hit parade in days to come. The second choice is the authorization stuff you went through earlier.

When you first start SPSS, you see a dialog box like the one shown in Figure 2-9. This dialog box lets you go directly to the window you want to work with. The problem is that it assumes you already know what you want to do, but you have no idea what you want to do with SPSS yet, so just click the Cancel button to close the dialog box.

You see the regular Data Editor window, shown in Figure 2-10. If you've ever worked with a spreadsheet, this display should look familiar, and it works much the same way. This window is the one you use to enter data. We generally like to expand the window to fill the entire screen because more spaces are displayed at one time. Besides, we don't need to see any other windows because we almost never do two things at once.

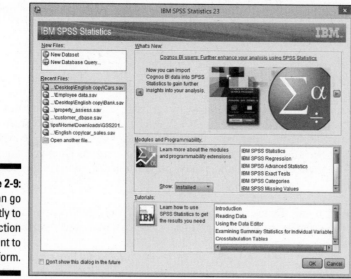

Figure 2-9:
You can go
directly to
the function
you want to
perform.

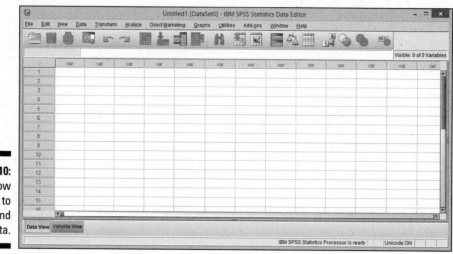

Figure 2-10:
The window
used to
enter and
view data.

Chapter 3

A Simple Statistical Analysis Example

The purpose of this chapter is to introduce you to the mechanics of working with SPSS. It begins with stepping through the process of entering some simple data into SPSS and continues with processing that data. This is followed by various procedures for deriving results, using a subset of the data for some calculations and other parts of the data for other calculations. Finally, the results from these different calculations are displayed in different ways.

The data for this example are simple, as are the displays that the data generate. The purpose of this chapter is not to present any great breakthrough in statistical analysis. Instead, we simply want to demonstrate the basic procedures you need to know about when you're using SPSS.

When the Tanana at Nenana Thaws

This analysis is about an annual lottery that takes place in Alaska. Actually, it isn't called a lottery — it's called a *classic,* whatever that means.

We don't know whether the Tanana Classic is the oldest lottery in the United States (it began in 1917), but it's certainly the slowest. It has only one jackpot per year, and tickets for that jackpot are sold all across the state during the winter months.

The lottery is simple enough: The citizens of the town of Nenana set up a large tripod on the ice in the middle of the Tanana River. From the top of the tripod, a tight line is stretched to a clock on a bridge. When the spring thaw comes, the tripod moves and the clock is triggered, stamping the exact minute. All the people who have selected the correct month, day, hour, and minute share the pot.

Many questions come to mind. What is the most likely date? What is the most likely time of day? Is there a trend? In the analysis that follows, we'll look at the answers to these questions and more.

By the way, the earliest the ice moved out was April 20 at 3:27 p.m. (in 1940), and the latest was May 20 at 11:41 a.m. (in 1964).

Entering the Data

SPSS can acquire data from many sources. You can instruct it to read data from a text file, a database, or a file produced by a program such as Microsoft Access or Microsoft Excel. This Alaskan example does it the simplest way possible: by typing data into the Data Editor window. (We said *simplest*, not easiest.)

The data consists of dates and times. SPSS has a special date format that we'll be using later, but for now, we'll enter the year, month, day, hour, and minute as separate numeric items. This keeps the example as simple as possible, and enables me to show you some different ways of manipulating numbers to reach conclusions.

Entering the data definitions

The first job is to define the names, labels, and data types for the various fields of data, also known as the *variables*. Here's all you need to do:

1. **Start the SPSS program by choosing Start⇨All Programs⇨IBM SPSS Statistics⇨SPSS Statistics 23.**

   Depending on how your software is configured, you may get an options window with OK and Cancel buttons. If so, click the Cancel button. In either case, an empty Data Editor window appears, as shown in Figure 3-1.

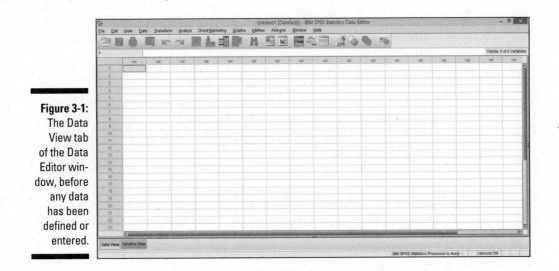

Figure 3-1:
The Data
View tab
of the Data
Editor win-
dow, before
any data
has been
defined or
entered.

The layout shown in Figure 3-1 is the Data View mode, as indicated by the tab at the bottom of the window. We want to go to the other mode.

2. Click the Variable View tab.

The window now looks like the one in Figure 3-2.

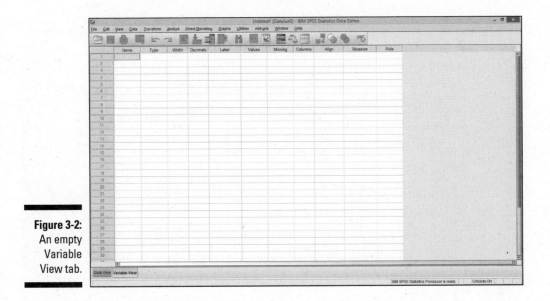

Figure 3-2:
An empty
Variable
View tab.

You use the Variable View tab to define the names and types of variables, and you use the Data View tab to enter the values for those variables.

To enter the definitions, you type the name in the first column — the one labeled Name — and then move the cursor down one row to the position for the next name in the list. You can most easily move the cursor by clicking the destination cell with the mouse. You can also move the cursor with the Enter key and the arrow keys, but the movement may not always be in the direction you expect.

In Figure 3-3, we entered the variable definitions we use in this example.

When you move down to define a new variable name, SPSS takes a wild guess at what you want in the cells you skipped and fills them in for you automatically. Some of the guesses are right, and some are wrong. Stick with us here, and we'll describe the fiddling around you have to do until your information matches that in Figure 3-3.

3. Type the following entries in the Name column:

- year

- month

- day

- hour

- minute

Every field has both a name and a label. One or the other is used as an identifying tag when data is displayed.

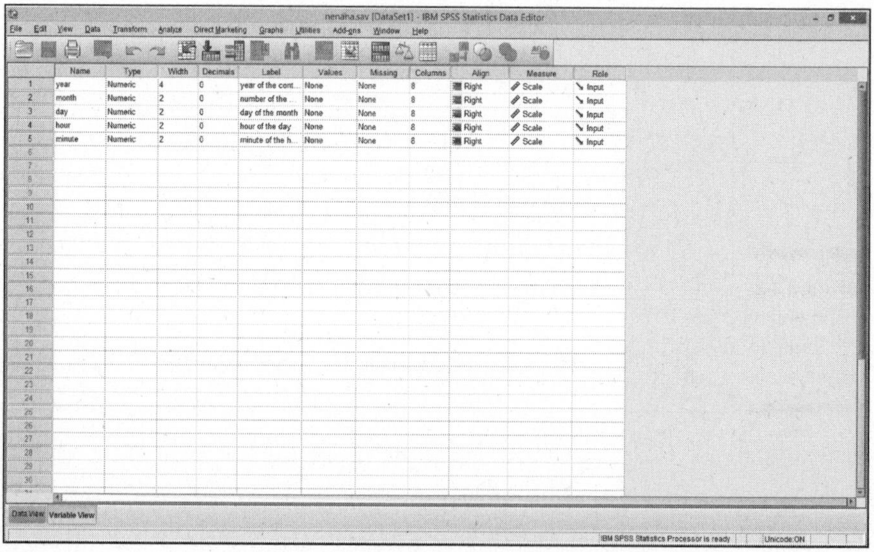

Figure 3-3:
Definition of the variable names.

The name is normally shorter than the label. A short name is handy when you're displaying data in a tight format (such as a column heading or a bar chart label) and when you're writing equations in the two scripting languages supplied with SPSS. The label is intended to be more descriptive and can add clarity by being displayed as descriptions in displays such as line graphs and pie charts.

4. Skip the Type column.

In this example, all the fields are simple numerics, so SPSS guesses correctly about most of the attributes and fills them in for you. Most of the data you enter into SPSS will be numeric, although some numbers will be converted into names by SPSS. It's hard to perform calculations with things like "moonbeam" and "sure bubba," but it can be done. Later on, we'll show you how to instruct SPSS to change numbers into words and phrases automatically.

SPSS has set the number of digits to the right of the decimal point (the Decimals column) to 2 for all the numbers in our example, but that's not what we want for this example.

5. Set all the values in the Decimals column to zero.

This has to be done before you can adjust the Widths — if you don't believe us, try it.

6. Set the first value in the Width column to 4 and the rest of the values in that column to 2.

The Width of most of the fields in this example should be changed from the default of 8 to 2 because they're two digits long. But set the year to a Width of 4 to accommodate four digits (we don't want to do the Y2K thing all over again). Simply click the box (cell) for the year's Width column and type **4**.

By the way, SPSS has a nifty Date data type. We didn't use it here because we want to show you how to work with simple numbers. You find out about dates and some other special types and formats in Chapter 4 and even more about dates in Chapter 7. At the end of this chapter, you get a sneak peak at the Date Time Wizard, something that we return to in Chapter 7.

7. Type the following into the cells in the Label column:

- year of the contest
- number of the month
- day of the month
- hour of the day
- minute of the hour

When you type in the Label column, you're not limited to the size of the cell that holds it. If you type a longer line, the box expands to take it all in. But don't write a thesis; you need something that will display nicely on your graphs and tables. (You can always come back and change it later on.)

Depending on how big you've made your window, you may have to scroll to display columns to either side. To scroll, use the horizontal scroll bar at the bottom of the screen. (We like to expand the window to the full screen, but that's probably because we're easily distracted if we see other windows.)

8. **Skip the Values column; it's for assigning names to specific values, and isn't used until later in this example.**

9. **In the column labeled Missing, specify whether it's okay to have values missing from this field.**

   For example, if you're taking a survey on what color underwear people are wearing, you could assign a number to each color, but you're bound to come across someone who's going commando, so you'll need to define a special value used to indicate a missing item. By default, SPSS doesn't allow for missing data, and this example doesn't have any, so the default is None.

10. **Skip the Columns column.**

    The default column width for a data item is 8, and that's okay for this example. You can make the columns smaller if you prefer, but you need to make sure the columns are big enough to hold your largest data item or its name. This is the amount of space that SPSS allocates when it constructs charts and tables. If you set the size too small, the data or the variable name will be cut for some displays.

11. **In the Align column, specify the alignment of your data.**

    You can choose whether the data should be aligned on the right, shoved over to the left, or placed in the center. Choose whatever you like. This is determined by personal preference, a lousy sense of design, and bad taste.

12. **Make sure the Measure column is set to Scale.**

    Scale is an amount or size — it's just a regular number — and works fine for what we're going to do. The other options are Ordinal and Nominal. Ordinal has to do with things that have a specific order. Nominal values are used to tag things as belonging to categories.

13. **Skip the Role column.**

    The Role of all variables in this example is standard: they hold input data. They could also be tagged as Target (or output) data, or as Both, or as None. A variable can be also designated as Partition and used to divide the data into separate samples.

Entering the actual data

Click the Data View tab, which is at the bottom of the Data Editor window, and the window changes to look like the one shown in Figure 3-4. The label names you entered in the Variable View tab appear at the top of the columns. This window is now ready for you to enter numeric data.

In Figure 3-4, notice the numbers down the left side of the window. This is the SPSS way of numbering rows, which are also called *cases.* If you use the scroll bar on the right side of the window to scroll down, you'll see these numbers change. You can think of these numbers as a road map to the layout in the window so you can keep track of where you are.

However, don't trust the numbers to identify your data. If you move your data from place to place in the grid, the numbers on the left don't move with it. That means if you insert a row, delete a row, or simply sort your data in a different way, the numbers on the left will associate with different sets of values — and your case numbers will all be different. If you need to identify a case in a manner that doesn't change when someone organizes the cases differently, you must add a field for identity and enter your own identifying numbers.

All the values that must be entered for this example are in the following list — but you can be lazy if you want to; we've already entered all the numbers. All you have to do is load the file that holds them by choosing File➪Open➪Data and then selecting nenana.sav. But even if you decide to read them in from the file, enter a few anyway so you can see how SPSS data entry works. (We talk about loading the file a little later.)

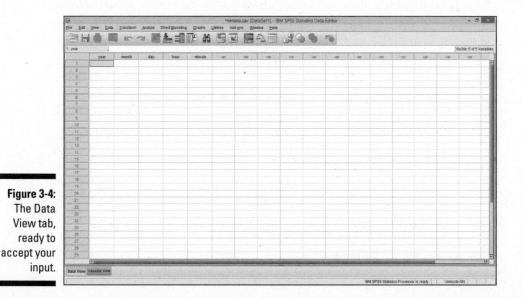

Figure 3-4:
The Data
View tab,
ready to
accept your
input.

4/20/1940, 15:27	5/1/1989, 20:14	5/8/1933, 19:30
4/20/1998, 16:54	5/1/1991, 12:04	5/8/1959, 11:26
4/23/1993, 13:01	5/1/2000, 10:47	5/8/1966, 12:11
4/24/1990, 17:19	5/2/1960, 19:12	5/8/1968, 21:26
4/24/2004, 14:16	5/2/1976, 10:51	5/8/1971, 21:31
4/26/1926, 16:03	5/2/2006, 17:29	5/8/1986, 22:50
4/26/1995, 13:22	5/3/1919, 14:33	5/8/2001, 13:00
4/27/1988, 09:15	5/3/1941, 13:50	5/9/1923, 14:00
4/28/1943, 19:22	5/3/1947, 17:53	5/9/1955, 14:13
4/28/1969, 12:28	5/4/1944, 14:08	5/9/1984, 15:33
4/28/2005, 12:01	5/4/1967, 11:55	5/10/1931, 09:23
4/29/1939, 13:26	5/4/1970, 10:37	5/10/1972, 11:56
4/29/1953, 15:54	5/4/1973, 11:59	5/10/1975, 13:49
4/29/1958, 14:56	5/5/1929, 15:41	5/10/1982, 17:36
4/29/1980, 13:16	5/5/1946, 16:40	5/11/1918, 09:33
4/29/1983, 18:37	5/5/1957, 09:30	5/11/1920, 10:46
4/29/1994, 23:01	5/5/1961, 11:30	5/11/1921, 06:42
4/29/1999, 21:47	5/5/1963, 18:25	5/11/1924, 15:10
4/29/2003, 18:22	5/5/1987, 15:11	5/11/1985, 14:36
4/30/1917, 11:30	5/5/1996, 12:32	5/12/1922, 13:20
4/30/1934, 14:07	5/6/1928, 16:25	5/12/1937, 20:04
4/30/1936, 12:58	5/6/1938, 20:14	5/12/1952, 17:04
4/30/1942, 13:28	5/6/1950, 16:14	5/12/1962, 21:23
4/30/1951, 17:54	5/6/1954, 18:01	5/13/1927, 05:42
4/30/1978, 15:18	5/6/1974, 15:44	5/13/1948, 11:13
4/30/1979, 18:16	5/6/1977, 12:46	5/14/1949, 23:39
4/30/1981, 18:44	5/7/1925, 18:32	5/14/1992, 06:26
4/30/1997, 10:28	5/7/1965, 19:01	5/15/1935, 13:32
5/1/1932, 10:15	5/7/2002, 20:27	5/16/1945, 09:41
5/1/1956, 23:24	5/8/1930, 19:03	5/20/1964, 11:41

You should now be seeing the Data View tab. To enter a number, simply click a position with the mouse and then type the number that you want to put in that square.

When we entered the data, we duplicated a row that was already there and then made changes to it. This was handy because the month and day of the new entry were often the same as the duplicated entry. To duplicate a row, select the row you want to copy by clicking the number at the left of the row. (One click selects the entire row.) Then choose Edit⇨Copy. Next, select the row you want to hold the duplicate data, and then choose Edit⇨Paste. If your target row already contains data, the new data overwrites it.

Suppose you want to insert a new row of data in front of some you already have. First, select the row that is in the position where you want to insert the new row; then choose Edit⇨Insert Cases to open a blank row in the position you've chosen. You can either copy or type new data into the blank row.

When you're finished, you can scroll up and down and see different parts of the data, as shown in Figure 3-5.

When you're entering your own data, select a filename early in the process and choose File⇨Save to write everything to that file from time to time. If you don't do this, a simple computer crash could cause you to lose all your data. That sort of thing is not good for your blood pressure.

Figure 3-5:
The data
freshly
entered into
SPSS.

	year	month	day	hour	minute
62	1969	5	8	11	26
63	1966	5	8	12	11
64	1968	5	8	21	26
65	1971	5	8	21	31
66	1986	5	8	22	50
67	2001	5	8	13	0
68	1923	5	9	14	0
69	1965	5	9	14	13
70	1964	5	9	15	33
71	1931	5	10	9	23
72	1972	5	10	11	56
73	1975	5	10	13	49
74	1962	5	10	17	36
75	1918	5	11	9	33
76	1920	5	11	10	46
77	1921	5	11	6	42
78	1924	5	11	15	10
79	1985	5	11	14	36
80	1922	5	12	13	20
81	1937	5	12	20	4
82	1952	5	12	17	4
83	1962	5	12	21	23
84	1927	5	13	5	42
85	1948	5	13	11	13
86	1949	5	14	23	39
87	1992	5	14	6	26
88	1936	5	15	13	32
89	1945	5	16	9	41
90	1964	5	20	11	41

By the way, if you've scrolled all the way down, you've noticed that there's a bottom to the list of numbered rows. Don't worry about it. As you enter data, the bottom extends so you never hit a limit.

If you've elected not to enter the data by hand, and instead you want to load it from the file, choose File⇨Open⇨Data, and then navigate to wherever you stored the `nenana.sav` file, as shown in Figure 3-6. Depending on how your Windows system is configured, the name may be chopped off in your display and appear only as `nenana`. It's not abnormal for Windows to change file-names this way. (The book's Introduction tells you how and where you can get the files.)

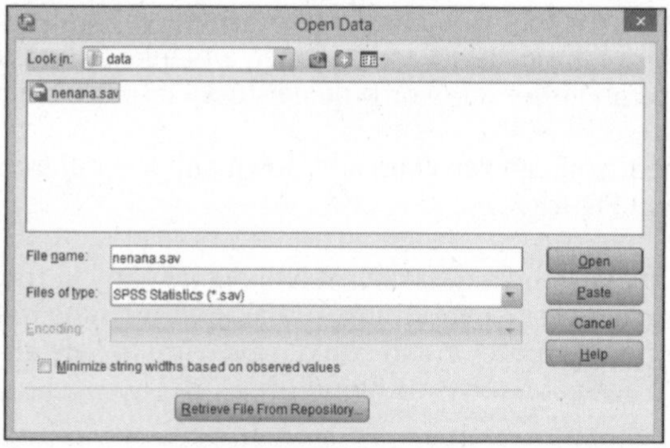

Figure 3-6:
Loading an
SPSS data
file.

The Most Likely Hour

After you've put the data in SPSS, do something simple. Use the following procedure to find the mean of the hours in an attempt to determine the hour of the day when the ice is most likely to melt. This would probably be in the daytime because the sun is warming both the air above the ice and the flowing water below the ice.

To find the most likely hour (ignoring the minutes for now), follow these steps:

1. **Choose Analyze⇨Descriptive Statistics⇨Descriptives.**

2. **In the box on the left, select** `hour of the day [hour]` **(one of your variable labels) and then click the arrow button in the middle of the window.**

   The label moves to the right, as shown in Figure 3-7.

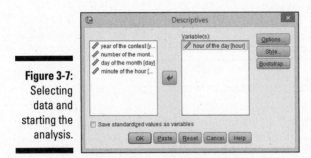

Figure 3-7:
Selecting
data and
starting the
analysis.

3. **Click the Options button.**

4. **Click the Mean, Std. deviation, Minimum, and Maximum check boxes, as shown in Figure 3-8.**

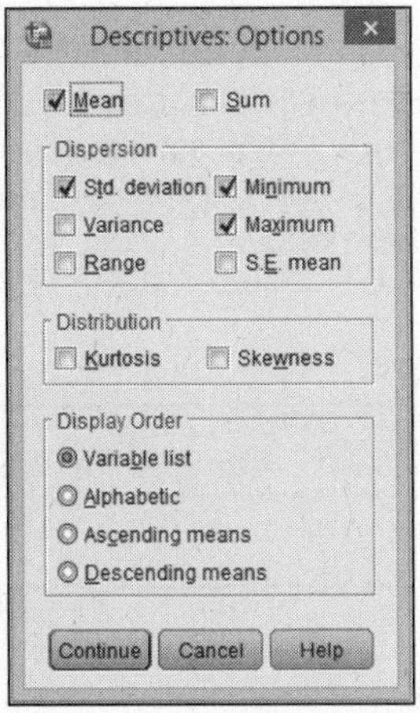

Figure 3-8:
The option
settings for
the analysis.

5. **Click Continue.**

6. **Click the OK button at the bottom-left of the window in Figure 3-7.**

The SPSS Statistics Viewer window appears and displays information about the analysis, including the results. A detailed description of all this information is in Chapter 8. For now, expand the window to fill the screen, and use the scroll bars if necessary, to locate the result in the box at the bottom of the right panel of the window, as shown in Figure 3-9. The mean (not the average, but nearly the same thing) shows the hour as 14.60, which is between 2 p.m. and 3 p.m. That makes sense, because that's near the warmest part of the spring day.

Descriptive Statistics

	N	Minimum	Maximum	Mean	Std. Deviation
hour of the day	90	5	23	14.60	4.069
Valid N (listwise)	90				

Inside the box, the text on the far left is the label you gave to the variable. The column labeled N is the number of data items included in the calculations. You can tell from the minimum and maximum that the earliest the ice has ever let go was during the 5 a.m. hour, but it has also been known to happen after 11 p.m.

The value for the standard deviation is calculated according to the degree of variation from a perfect fit on a bell curve.

There's more bell-curve stuff to diddle with: Go back through the same procedure again, but this time change the options in Step 4. Remove maximum and minimum and instead enable Kurtosis and Skewness. Those are not rude words (and, no, we didn't just make them up); they're types of statistics. As shown in Figure 3-10, the results have two new values.

Descriptive Statistics

	N	Mean	Std. Deviation	Skewness		Kurtosis	
	Statistic	Statistic	Statistic	Statistic	Std. Error	Statistic	Std. Error
hour of the day	90	14.60	4.069	.086	.254	-.480	.503
Valid N (listwise)	90						

Both values also have to do with the bell curve. *Skewness* represents the symmetry of the data. A positive skewness indicates that more of the data appears to the high end, or the right, on the graph. A negative value indicates a skew to the lower values. *Kurtosis* has to do with the flatness of the curve. If the data implies a curve flatter than the bell curve, the kurtosis value is negative. If, on the other hand, the data inscribes a curve that is more pointed on top than the bell curve, the kurtosis value is positive.

Transforming Data

The previous example looks at only the hours, but it's also possible to include minutes. Clock arithmetic is tricky (it's that 60-minutes-per-hour thing) but SPSS can work with it if you tell it what you're doing.

In the next example, we combine the separate hour and minutes fields into a new field that contains both. SPSS is good at transforming data this way. To build the new field, do the following:

1. In the Data Editor window, choose Transform⇨Date and Time Wizard.

The window shown in Figure 3-11 appears.

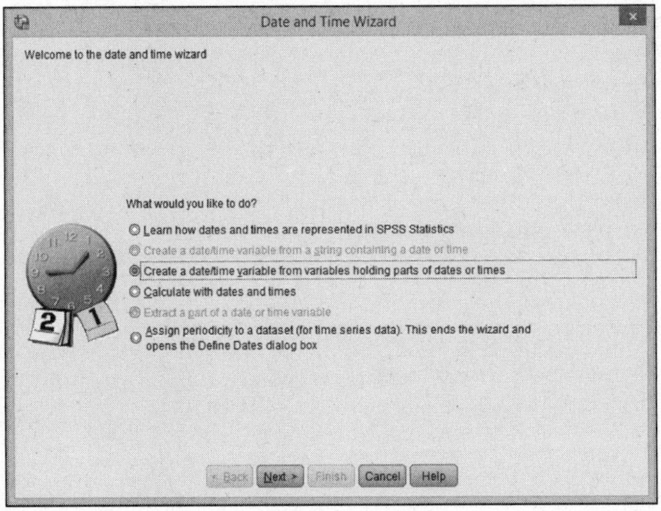

Figure 3-11:
The Date and Time Wizard.

2. Select the Create a Date/Time Variable from Variables Holding Parts of Dates or Times option.

3. Click Next.

4. Put the names of the variables into the appropriate fields.

We want only the hours and minutes, so ignore the others. You move them by selecting the one you want from the list on the left, and then clicking the arrow next to the place you want it to go. When you're finished, the screen should look like Figure 3-12.

5. Click Next.

6. Enter a name and a label for the variable. Also select a display format from the list.

To follow along with the example, type **time** in the Result Variable box, type **hour and minute** in the Variable Label box, and then select hh:mm from the Output Format list, as shown in Figure 3-13.

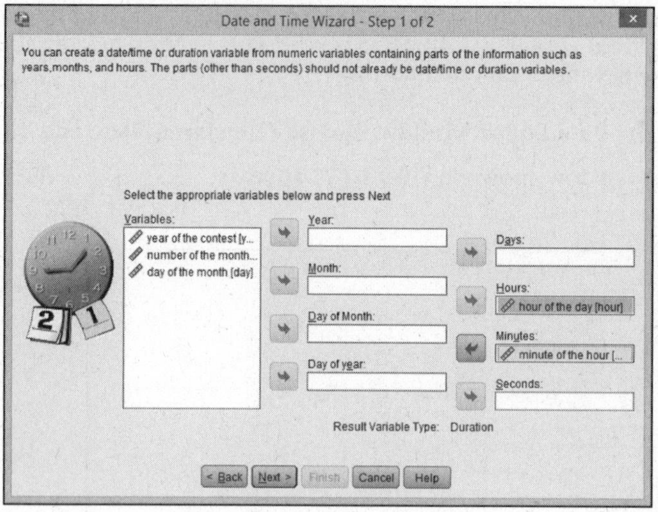

Figure 3-12: Selection of the variables from which time is structured.

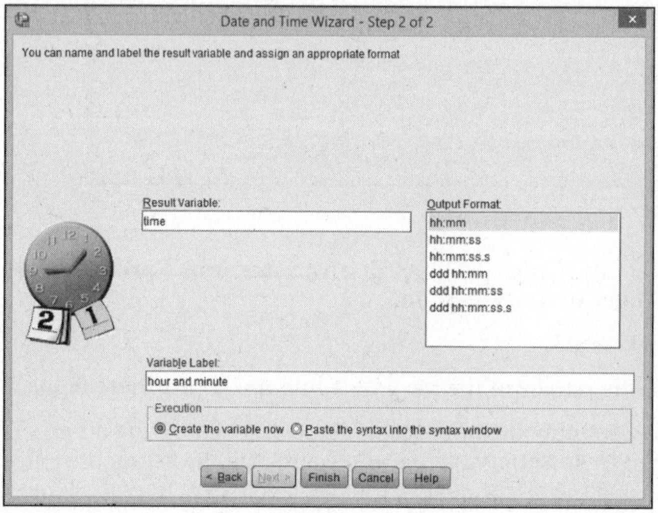

Figure 3-13: The name and display format for times.

7. **Select the Create the Variable Now option, and then click the Finish button.**

You've created your new time data field. The result is shown in Figure 3-14.

You may notice that after you click the Finish button, the SPSS Statistics Viewer window may appear. This window shows a log of the steps the program has performed. To continue, click back into the Data Editor window.

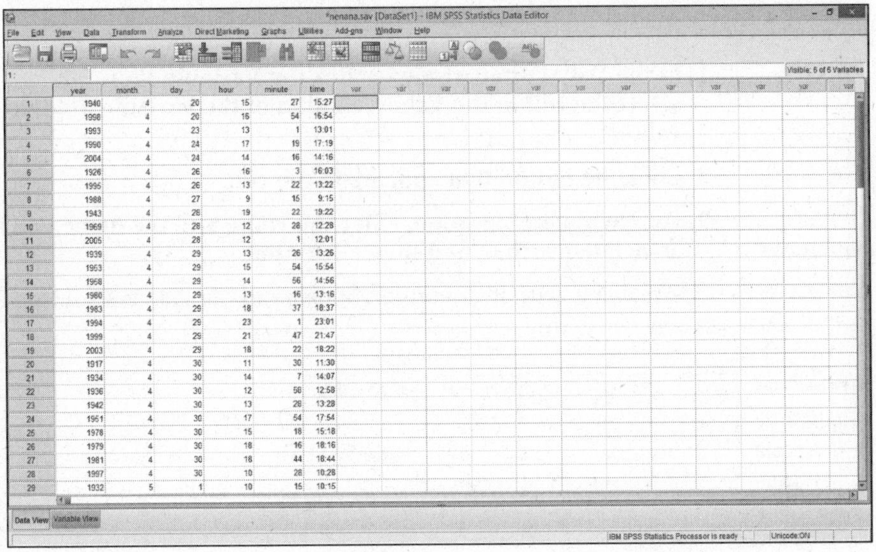

	year	month	day	hour	minute	time
1	1940	4	20	15	27	15:27
2	1998	4	20	16	54	16:54
3	1993	4	23	13	1	13:01
4	1990	4	24	17	19	17:19
5	2004	4	24	14	16	14:16
6	1926	4	26	16	3	16:03
7	1995	4	26	13	22	13:22
8	1988	4	27	9	15	9:15
9	1943	4	28	19	22	19:22
10	1969	4	28	12	28	12:28
11	2005	4	28	12	1	12:01
12	1939	4	29	13	26	13:26
13	1953	4	29	15	54	15:54
14	1958	4	29	14	56	14:56
15	1980	4	29	13	16	13:16
16	1983	4	29	18	37	18:37
17	1994	4	29	23	1	23:01
18	1999	4	29	21	47	21:47
19	2003	4	29	18	22	18:22
20	1917	4	30	11	30	11:30
21	1934	4	30	14	7	14:07
22	1936	4	30	12	58	12:58
23	1942	4	30	13	28	13:28
24	1951	4	30	17	54	17:54
25	1978	4	30	15	18	15:18
26	1979	4	30	18	16	18:16
27	1981	4	30	18	44	18:44
28	1997	4	30	10	28	10:28
29	1932	5	1	10	15	10:15

Figure 3-14: The Data Editor window with the new time field.

Now follow the same procedure as before by choosing Analyze➪Descriptive Statistics➪Descriptives. But in Step 2, select only the new time field so you can see how SPSS handles different combinations of values. In the results, look at the difference in the two means: When the minutes are included, the mean moves to a time a bit later (as one would expect). It's now at 15:03 (3:03 p.m.) instead of 2-something. Whether that difference is statistically significant is up to you.

The Two Kinds of Numbers

With this example data so far, we've dealt with *continuous variables* (amounts and distances, such as age, gallons of gas, and the number of beans in a jar). The other type is *categorical variables* (where each value represents a category — for example, yes and no [where, for example, yes is 1 and no is 0] and types of balls [where 1 is a football, 2 is a soccer ball, 3 is a snooker ball, and so on]).

All the variables in this example — except the number indicating the month — are continuous variables. We tend to think of the months by their names instead of numbers, but you have to use the number of the month to do any calculations. If you want the name displayed, you have to assign a descriptive name for each possible value. That's easy to do with this data because we have only two values: 4 and 5.

To add identifiers for the values, do the following:

1. **In the Data Editor window, click the Variable View tab and then select the cell in the Values column of the variable holding the month values.**

2. **Click the button that appears in the cell.**

3. **For each possible value, enter the value and the name you want associated with it, and then click Add.**

   The value, with its identifier, appears in the list, as shown in Figure 3-15.

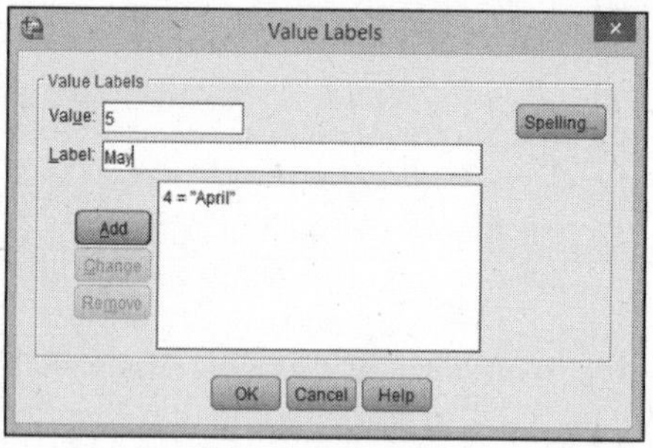

Figure 3-15: One name has been added for a value and another one is being entered.

4. **After you've added all the values you want to define, click OK.**

   The screen displays only part of the change. The word None is gone, and in its place is part of one of your new value definitions.

The real result will show up in your output and help you make a lot more sense of your results. For example:

1. **Choose Graphs⇨Legacy Dialogs⇨Pie.**

   The window shown in Figure 3-16 appears.

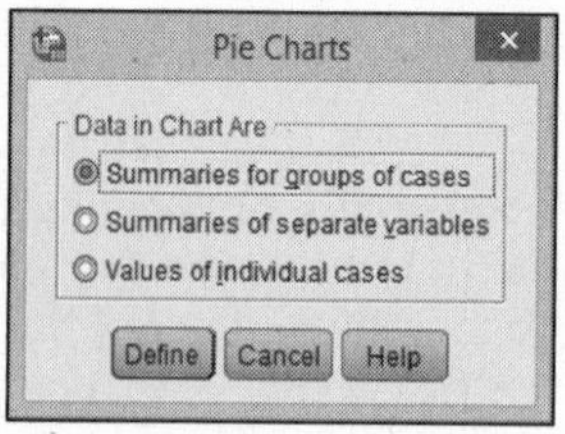

Figure 3-16: Select the type of data to be displayed in the pie chart.

2. **Select the Summaries for Groups of Cases option, and then click the Define button.**

3. **In the column on the left, select** number of the month [month], **and then click the arrow to the left of Define Slices By, as shown in Figure 3-17.**

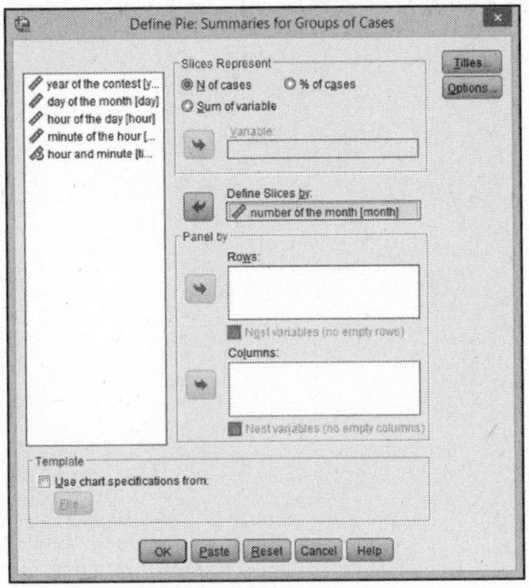

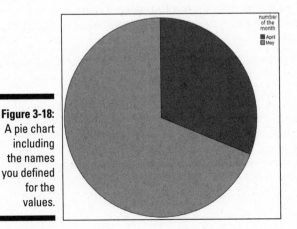

Figure 3-17: You can select the variables you want for the pie divisions.

4. **Click OK.**

The SPSS Statistics Viewer window appears, as shown in Figure 3-18.

Figure 3-18: A pie chart including the names you defined for the values.

The Day It's Most Likely to Happen

You already know that the ice is most likely to move in the warmer part of the day. A quick graph can show you whether there's a most likely day as well. To get a quick bar graph, do the following:

1. **Choose Graphs⇨Legacy Dialogs⇨Bar.**

   The dialog box shown in Figure 3-19 appears.

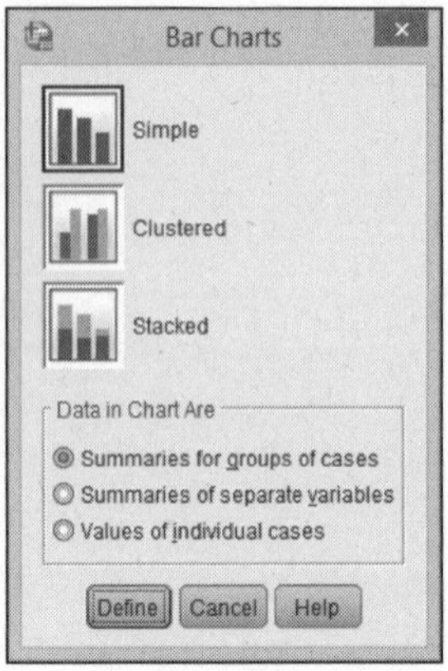

Figure 3-19:
You can select the fundamentals of the bar chart you want.

2. **Select the Simple bar chart and the Summaries for Groups of Cases option, and then click the Define button.**

3. **For Bars Represent, select N of Cases, which means the bars will represent the number of cases. Also set the Category Axis to** day of the month [day] **and set the Rows to** number of the month [month]**, as shown in Figure 3-20.**

   The exact meanings of these terms and settings are explained in Part IV, which covers graphs.

4. **Click OK.**

   The bar chart shown in Figure 3-21 appears. The chart shows which days in the past were most often the ones on which the ice moved. There is no obvious trend that we can see. However, you may want to experiment with different analysis displays and try to find a pattern.

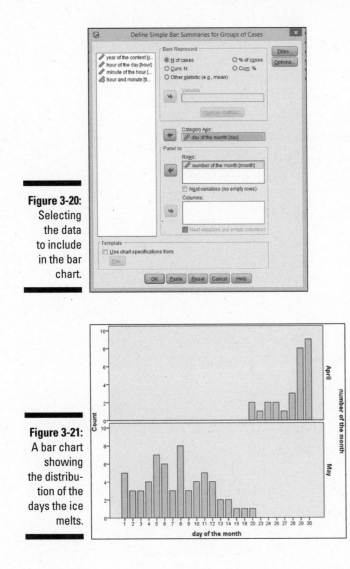

Figure 3-20: Selecting the data to include in the bar chart.

Figure 3-21: A bar chart showing the distribution of the days the ice melts.

Part II
Getting Data in and out of SPSS

Variable Type   ✕

◉ Numeric

○ Comma   Width: [8]

○ Dot   Decimal Places: [2]

○ Scientific notation

○ Date

○ Dollar

○ Custom currency

○ String

○ Restricted Numeric (integer with leading zeros)

ⓘ The Numeric type honors the digit grouping setting, while the Restricted
Numeric never uses digit grouping.

[OK] [Cancel] [Help]

Find out about the Automatic Recode command in a free article at www.dummies.com/
extras/spss.

In this part . . .

- ✔ Get set up in a new dataset like a pro.
- ✔ Get data into SPSS.
- ✔ Get results out of SPSS.
- ✔ Master all of the SPSS data types.

Chapter 4

Entering and Defining Data

. . . .

In This Chapter

▶ Considering your choices when defining a variable

▶ Defining variables

▶ Entering numbers

▶ Making sure that you're using the right measurement type

. . . .

*T*o process your data, you have to get it into the computer. Entering data has been a problem with computers since the beginning. No matter how you decide to get your numbers into SPSS, at some point someone has to type them (unless they come from some form of automatic monitoring). These days, it feels like we spend half of our time entering data into online forms, which saves some analyst from typing on the other end. SPSS can read data from other places. You can also type directly into SPSS — and, if you want, copy the data to places other than SPSS later.

Entering data into SPSS is a two-step process: First, you define what sort of data you'll be entering. Then you enter the actual numbers. This may sound difficult, but it isn't so bad. When you see how data entry works in SPSS, you'll discover you have some pretty nifty software to help you.

You organize your data into cases. Each case is made up of a collection of variables. First, you define the characteristics of the variables that make up a case, and then you enter the data into the variables to make up the contents of the cases. This chapter shows you how to work with this technique of getting data into your system.

Entering Variable Definitions on the Variable View Tab

You use the Variable View tab of the Data Editor window, shown in Figure 4-1, to define the names and characteristics of variables. This is where you always start if you plan on entering data into SPSS. As you can see in Figure 4-1, every

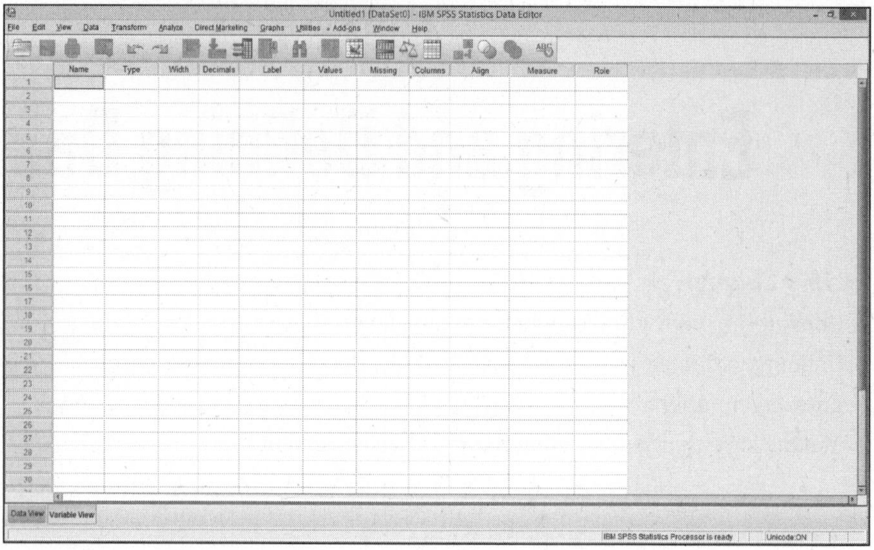

Figure 4-1:
You use the
Variable
View tab to
define the
charac-
teristics of
variables.

characteristic you can define about your variables is named at the top of
the window. All you have to do is enter something in each column for each
variable.

The predefined set of 11 characteristics are the only ones needed to
completely specify all the attributes of any variable. The characteristics are
all known to the internal SPSS processing. When you add a new variable,
you'll find that reasonable defaults appear for most characteristics.

The Variable View tab is just for defining the variables. The entry of the
actual numbers comes later (see "Entering and Viewing Data Items on the
Data View Tab," later in this chapter).

Each variable characteristic has a default, so if you don't specify a characteristic, SPSS fills one in for you. However, what it selects may not be what you
want, so let's look at all the possibilities.

Name

The cell on the far left is where you enter the name of the variable. Just click
the cell and type a short descriptor, such as **age**, **income**, **sex**, or **odor**. (A
longer descriptor, called a _label,_ comes later.) You can type longer names
here, but you should keep them short because they'll be used in named lists
and as identifier tags on the data graphs and such — where the format can be
a bit crowded. Names that are too long can cause the output from SPSS to be
garbled or truncated.

If the name you assigned turns out to be too long or is misspelled, you can always change it on the Variable View tab. One of the nice things about SPSS is that you can correct mistakes quickly.

Here are some handy hints about names:

✔ **You can use some bizarre characters in a name, such as @, #, and $, as well as the underscore character (_) and numbers.** But if you use screwy characters in a name, you may live to regret it. For one thing, you can't start a variable name with these characters. Plus, they'll remind expert users of special variables in some advanced features. An underscore in the middle of a name is a great way to make a name more readable, but otherwise, it's best to keep your names simple.

✔ **Be sure to start every name with an uppercase or lowercase letter.**

✔ **You can't include blanks anywhere in a name, but an underscore is a good substitute.**

If you want to export data to another application, make sure the names you use are in a form acceptable to that application. Watch out for special characters.

Type

Most data you enter will be just regular numbers. Some, however, will be a special type, such as currency, and some will be displayed in a special format. Other data, such as dates, will require special procedures for calculation. You simply specify what type you have, and SPSS takes care of those other details for you. This is a comprehensive look at all the types. (We give you more advice about some special types in Chapter 7.)

Click the cell in the Type column you want to fill in, and a button with three dots appears on its right. Click that button, and the Variable Type dialog box, shown in Figure 4-2, appears.

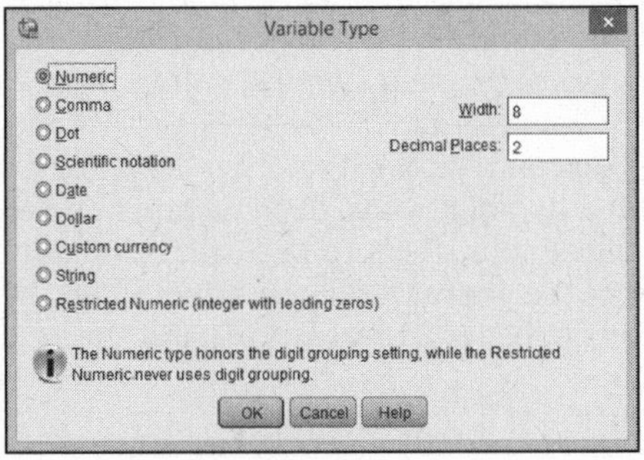

Figure 4-2:
The Variable Type dialog box allows you to specify the type of variable you're defining.

Variable Type

○ Numeric
○ Comma
○ Dot
○ Scientific notation
○ Date
○ Dollar
○ Custom currency
○ String
○ Restricted Numeric (integer with leading zeros)

Width: 8
Decimal Places: 2

The Numeric type honors the digit grouping setting, while the Restricted Numeric never uses digit grouping.

OK   Cancel   Help

You can choose from the following predefined types of variables:

- **Numeric:** Standard numbers in any recognizable form. The values are entered and displayed in the standard form, with or without decimal points. Values can be formatted in standard scientific notation, with an embedded E to represent the start of the exponent. The Width value is the total number of all characters in a number — including any positive or negative signs and the exponent indicator. The Decimal Places value specifies the number of digits displayed to the right of the decimal point, not including the exponent.

- **Comma:** This type specifies numeric values with commas inserted between three-digit groups. The format includes a period as a decimal point. The Width value is the total width of the number, including all commas and the decimal point. The Decimal Places value specifies the number of digits to the right of the decimal point. You may enter data without the commas, but SPSS will insert them when it displays the value. Commas are never placed to the right of the decimal point.

- **Dot:** Same as Comma, except a period is used to group the digits into threes, and a comma is used for the decimal point.

- **Scientific Notation:** A numeric variable that always includes the E to designate the power-of-ten exponent. The *base* (the part of the number to the left of the E) may or may not contain a decimal point. The *exponent* (the part of the number to the right of the E, which also may or may not contain a decimal) indicates how many times 10 multiplies itself, after which it's multiplied by the base to produce the actual number. You may enter D or E to mark the exponent, but SPSS always displays the number using E. For example, the number 5,286 can be written as 5.286E3. To represent a small number, the exponent can be negative. For example, the number 0.0005 can be written as 5E–4. This format is useful for very large or very small numbers.

- **Date:** A variable that can include the year, month, day, hour, minute, and second. When you select Date, the available format choices appear in a list on the right side of the dialog box, as shown in Figure 4-3. Choose the format that best fits your data. Your selection determines how SPSS will format the contents of the variable for display. This format also determines, to some extent, the form in which you enter the data. You can enter the data using slashes, colons, spaces, or other characters. The rules are loose — if SPSS doesn't understand what you enter, it tells you, and you can re-enter it another way. For example, if you select a format with a two-digit year, SPSS accepts and displays the year that way, but it will use four digits to perform calculations. The first two digits (the number of the century) will be selected according to the configuration you set by choosing Edit⇨Options and then clicking the Data tab.

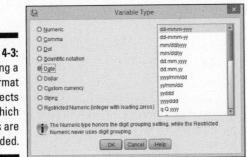

Figure 4-3:
Selecting a
date format
also selects
which
items are
included.

✔ **Dollar:** When you select Dollar, the available format choices appear in
a list on the right side of the dialog box (see Figure 4-4). Dollar values
are always displayed with a leading dollar sign and a period for a decimal point; for large values, they include commas to collect the digits
in groups of threes. You select the format and its Width and Decimal
Places values. The format choices are similar, but it's important that you
choose one that's compatible with your other dollar-variable definitions
so they line up when you print and display monetary values in output
tables. The Width and Decimal Places settings help with vertical alignment in the output, no matter how many digits you include in the format
itself. No matter what format you choose, you can enter the values without the dollar sign and the commas; SPSS inserts those for you.

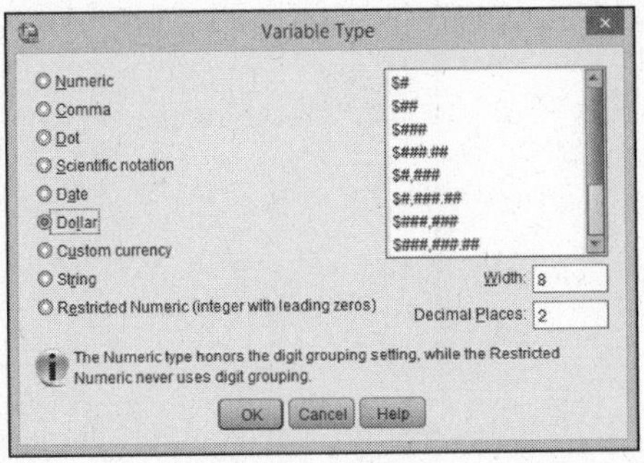

Figure 4-4:
The different
dollar for-
mats mostly
specify the
number of
digits to be
included.

✔ **Custom Currency:** The five custom formats for currency are named
CCA, CCB, CCC, CCD, and CCE, as shown in Figure 4-5. You can view and
modify the details of these formats by choosing Edit⇨Options and then
clicking the Currency tab. Fortunately, you can modify the definitions of
these custom formats as often as you like without fear of damaging your
data. As with the Dollar format, the Width and Decimal Places settings
are primarily for aligning the data when you're printing a report.

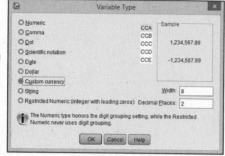

Figure 4-5:
Five custom
currency
formats are
available.

✔ **String:** A freeform non-numeric item (see Figure 4-6). The only good time to use string is when it truly is a string, like an address, a proper name, or a product code (SKU). Avoid using the String type when it really should be labeled Numeric. Something like favorite color, sex, or state should *not* be a string because it has a finite list of possibilities that are known in advance. (See the "Values" section later in this chapter.)

SPSS allows a very large number for the size of the string — so large that you could fit a paragraph, which is exactly what you would do if you were doing text mining. Open-ended response items in a survey would also be an example of a string.

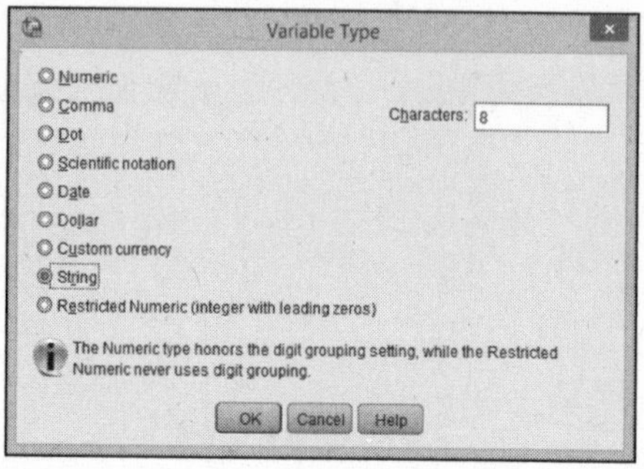

Figure 4-6:
Strings are
text like
addresses,
names, and
open-ended
responses.

✔ **Restricted Numeric:** A relatively new choice, so you may not see it mentioned in older books about SPSS. This is perfect for numbers that sometimes have leading zeros like zip codes and Social Security numbers. They aren't really numbers because you don't perform arithmetic on them. Back in the day, these types of numbers had to be declared as strings.

Width

The width setting in the definition of a variable determines the number of characters used to display the value. If the value to be displayed is not large enough to fill the space, the output will be padded with blanks. If it's larger than you specify, it will either be reformatted to fit or asterisks will be displayed.

Certain type definitions allow you to set a width value. The width value you enter as the width definition is the same as the one you enter when you define the type. If you make a change to the value in one place, SPSS changes the value in the other place automatically. The two values are the same.

At this point, you can do one of three things:

- ✔ Skip this cell and accept the default (or the number you entered previously under Type).
- ✔ Enter a number and move on.
- ✔ Use the up and down arrows that appear in the cell to select a numeric value.

Decimals

The number of decimals is the number of digits that appear to the right of the decimal point when the value appears onscreen. This is the same number that you may have specified as the Decimal Places value when you defined the variable type. If you entered a number there, it appears here as the default. If you enter a number here, it changes the one you entered for the type. They're the same.

Now you can do one of three things:

- ✔ Skip this cell and accept the default (or the number you entered earlier under Type).
- ✔ Enter a number and move on.
- ✔ Use the up and down arrows that appear in the cell to select a numeric value.

Label

The name and the label serve the same basic purpose: They're descriptors that identify the variable. The difference is that the *name* is the short identifier and the *label* is the long one. You need one of each because some output formats work fine with a long identifier and other formats need the short form.

You can use just about anything for the label. What you choose has to do with how you expect to use your data and what you want your output to look like. For example, the variable name may be "Sex" and the longer label may be "Boys and Girls," "Men and Women," or simply "Gender."

The length of the label isn't determined by some sort of software requirement. However, output looks better if you use short names and somewhat longer labels. Each one should make sense standing alone. After you produce some output, you may find that your label is lousy for your purposes. That's okay; it's easy to change. Just pop back to the Variable View tab and make the change. The next time you produce output, the new label will be used.

You can also just skip defining a label. If you don't have a label defined for a variable, SPSS will use the name you defined for everything.

Values

The Values column is where you assign labels to all the possible values of a variable. If you select a cell in the Values column, a button with three dots appears. Clicking that button displays the dialog box shown in Figure 4-7.

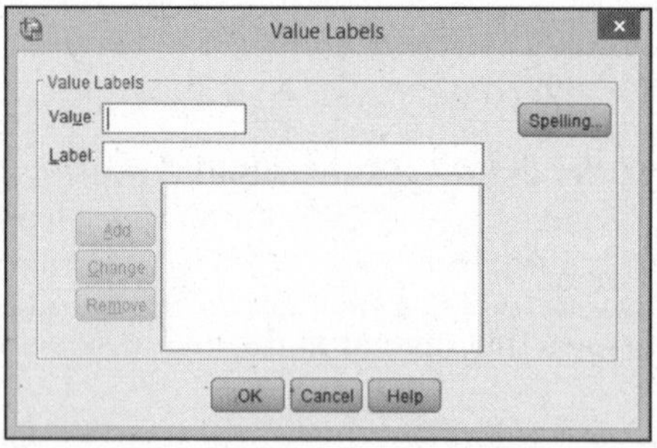

Figure 4-7:
You can assign a name to each possible value of a variable.

Value Labels
Value Labels
Value:
Label:
Spelling
Add
Change
Remove
OK   Cancel   Help

Normally, you make one entry for each possible value that a variable can assume. For example, for a variable named Sex you could have the value 1 assigned the label "Male" and 2 assigned the label "Female." Or, for a variable named Committed you could have 0 for "No," 1 for "Yes," and 2 for "Undecided." If you have labels defined, when SPSS displays output, it can show the labels instead of the values.

To define a label for a value:

1. **In the Value box, enter the value.**
2. **In the Label box, enter a label.**
3. **Click the Add button.**

   The value and label appear in the large text block.

4. **To change or remove a definition, simply select it in the text block and make your changes; then click the Change button.**
5. **Repeat Steps 1–4 as needed.**
6. **Click OK to save the value labels and close the dialog box.**

You can always come back and change the definitions using the same process you used to enter them. The dialog box will reappear, filled in with all the definitions; then you can update the list.

Sometimes you have a whole bunch of strings and you really don't want to make them all values because it seems like it'll be a lot of work. A variable like college major is a good example. If you dread setting up 1 as "Astrophysics," 2 as "Biology," 3 as "Chemistry," and so on, you can use a special dialog box called Automatic Recode (under the Transform menu) and it'll do all the work for you.

Missing

You can specify what is to be entered for a value that is missing for a variable in a case. In other words, when you have values for all variables in a case except one, you can specify a placeholder for the missing value. Select a cell in the Missing column. Click the button with three dots and the Missing Values dialog box, shown in Figure 4-8, appears.

For example, say you're entering responses to questions, and one of the questions is, "How many cars do you own?" The normal answer to this question is a number, so you define the variable type as a number. If someone chooses to ignore this question, this variable won't have a value. However,

Figure 4-8:
You can
specify
exactly
what is
entered for
a missing
value.

you can specify a placeholder value. Perhaps 0 seems like a good choice for a placeholder here, but it's not really — lots of people don't have cars. Instead, a less likely value — like, say, -1 — makes a better choice. A very popular choice among SPSS users is -9, but this will depend on the values of the original variable.

You can even specify unique values to represent different reasons for a value being missing. In the previous example, you could define -1 as the value entered when the answer is, "I don't remember," and -2 could be used when the answer is, "None of your business." If you specify that a value is representing a missing value, that value is not included in general calculations. During your analysis, however, you can determine how many values are missing for each of the different reasons. You can specify up to three specific values (called *discrete values*) to represent missing data, or you can specify a range of numbers along with one discrete value, all to be considered missing. The only reason you would need to specify a range of values is if you have lots of reasons why data is missing and want to track them all.

One of the many reasons you don't want to abuse the string type is that it makes a mess of missing data or incorrect data. If Female and Male are strings, you can get entries like "m," "M," "Male," and even crazy unexpected ones like "H" and "mail." You're better off doing what all experienced users do: Use numeric codes with values!

Columns

The Columns column is where you specify the width of the column you'll use to enter the data. The folks at SPSS could have used the word *Width* to describe it, but they already used that term for the width of the data itself. A better name may have been the two words *Column Width,* but that would have been too long to display nicely in this window, so they just called it *Columns*. To specify the number of columns, select a cell and enter the number.

Align

The Align column determines the position of the data in its allocated space, whenever the data is displayed for input or output. The data can be left-aligned, right-aligned, or centered. You've defined the width of the data and the size of the column in which the data will be displayed; the alignment determines what is done with any space left over.

When you select a cell in the Align column, a list appears and you can choose one of the three alignment possibilities, as shown in Figure 4-9. Aligning to the left means inserting all blanks on the right; aligning to the right inserts all the extra spaces on the left; centering the data splits the spaces evenly on each side — we don't know what it does if an odd space is left over. (We also worry about things like the number of seeds in a tomato and where the clouds go at night.)

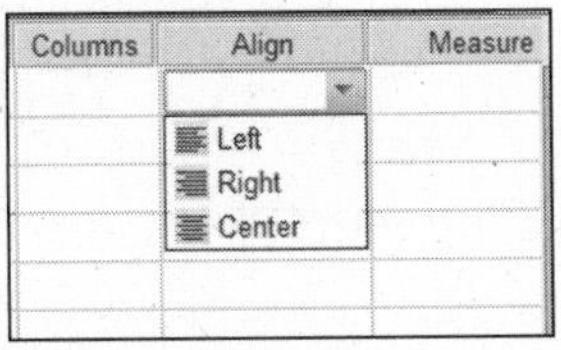

Figure 4-9:
Values can be justified right or left, or posi-tioned in the center.

Columns	Align	Measure
	Left	
	Right	
	Center	

Measure

Your value here specifies the measure of something in one of three ways. When you click a cell in the Measure column, you can select one of these choices (see Figure 4-10):

- ✔ **Ordinal:** These numbers specify the position (order) of something in a list. For example, *first, second,* and *third* are ordinal numbers.

- ✔ **Nominal:** Numbers that specify categories or types of things. You can have 0 represent "Disapprove" and 1 represent "Approve." Or you can use 1 to mean "Fast" and 2 to mean "Slow."

- ✔ **Scale:** A number that specifies a magnitude. It can be distance, weight, age, or a count of something.

Figure 4-10:
The type of measure-
ment being made by the
values in this variable.

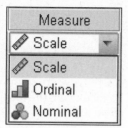

Role

Some of the SPSS dialog boxes select variables according to their role and include them as defaults. You don't need to worry about this characteristic. It can be handy when you have some experience with SPSS and understand how defaults are chosen.

Be on the lookout for Analyze menus that let you use predefined roles. They use this feature. There aren't that many of them, but the number grows with each new version of SPSS.

These predefined roles allow greater capability with SPSS Modeler, which is a kind of sibling product to SPSS Statistics.

When you click a cell in the Role column, you can select one of six choices (see Figure 4-11):

Figure 4-11:
The role assumed by this variable in certain SPSS dialog boxes.

✔ **Input:** This variable is used for input. This is the default role. Definition of roles was introduced to Version 18 of SPSS, and all data imported from earlier versions will be assigned this role.

✔ **Target:** This variable is used as output by SPSS procedures.

✔ **Both:** This variable is used as both input and output.

✔ **None:** This variable has no role assignment.

✔ **Partition:** This variable is used to partition the data into separate samples for training, testing, and validation.

✔ **Split:** This variable is used to build separate models for each possible value of the variable. This capability should not be confused with file splitting (see Chapter 8).

Entering and Viewing Data Items on the Data View Tab

After you've defined all the variables for each case, click the Data View tab of the Data Editor window so you can begin typing the data. At the top of the columns in Figure 4-12, you can see some names we chose for variables. Switching to the Data View tab makes the window ready to receive entered data — and to verify that what's entered matches the specified format and type of the data.

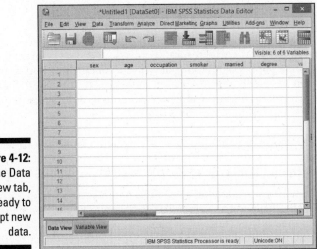

Figure 4-12: The Data View tab, ready to accept new data.

Entering data into one of these cells is straightforward: You simply click the cell and start typing.

If something is already in a cell and you want to change it instead of just typing over it, look up toward the top of the window, just underneath the toolbar: You'll see the name of the variable and the currently selected value. Click the value in the field at the top, and you can edit it right there. You can do all the normal mouse and keyboard stuff there, too — you can use the Backspace key to erase characters, or select the entire value and type right over it.

If you feel like a lousy (or inexperienced) mouse driver, take some time to experiment and figure out how to edit data. Lots of software use these same editing techniques, so becoming proficient now will pay you dividends later.

If your data is already in a file, you may be able to avoid typing it in again by reading that file directly into SPSS. For more information, see Chapter 5.

Don't take chances. As soon as you type a few values, save your data to a file by choosing File⇨Save As. Then choose File⇨Save throughout the process of entering data, and you won't be ruined if the computer crashes unexpectedly.

We all have to go back and refine our variable definitions from time to time. That's normal. When you come across something that doesn't do what you want it to, just switch back to the Variable View tab and correct it. Nobody but you and SPSS will ever know about it, and SPSS never talks.

Filling In Missed Categorical Values

Now that you've defined your variables and entered your data, you may want to check that you have names defined for all your actual ordinal and nominal values, and that you have defined the correct measures for them. SPSS can help by scanning your data, finding values for which you don't have definitions, and pointing them out in a friendly way.

The following steps use an existing file to walk through a demonstration:

1. **Choose File⇨Open⇨Data to load the file named** `car_sales.sav`.

   This file came with your installation of SPSS and is found, along with a number of other files, in the same directory in which you installed SPSS. You can load any of these data files, but `car_sales.sav` is the one used in this demonstration. If you load this file while you already have some other data showing in the window, SPSS will open a new Data Editor window to display the new information; your existing data will not be lost.

   When you open this data file — or any data file, for that matter — SPSS opens the SPSS Statistics Viewer window to tell you that it has opened a file (or the information could be displayed in the SPSS Statistics Viewer window that's already open). You won't need this information for what you're doing here, so you can just close the window.

2. **Choose Data⇨Define Variable Properties.**

   The Define Variable Properties dialog box appears.

3. **On the left, select all the names of the variables you want to check, and then click the arrow in the center of the dialog box to move them to the right, as shown in Figure 4-13.**

4. **When you're done, click Continue.**

5. **Select one of the variable names in the list on the left.**

   Its different values appear in the center of the dialog box, as shown in Figure 4-14. (In this example, every value has a name assigned to it.)

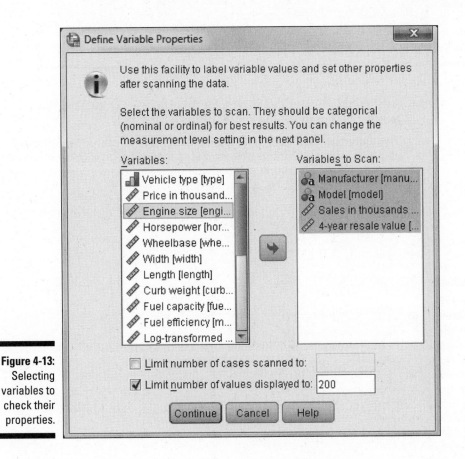

Figure 4-13:
Selecting variables to check their properties.

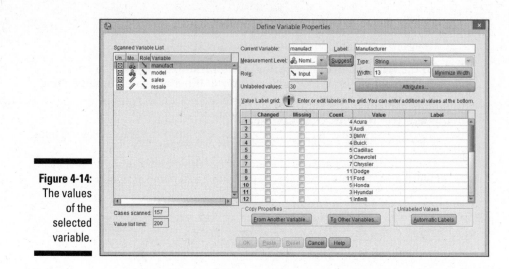

Figure 4-14:
The values
of the
selected
variable.

6. **Ask SPSS to suggest a new type for this variable by clicking the Suggest button in the top center of the dialog box.**

The dialog box in Figure 4-15 appears, telling you what SPSS concludes about this variable and its values. This same window, with different text, appears for each variable you test. Sometimes the text suggests changes in the variable definition, and sometimes it doesn't.

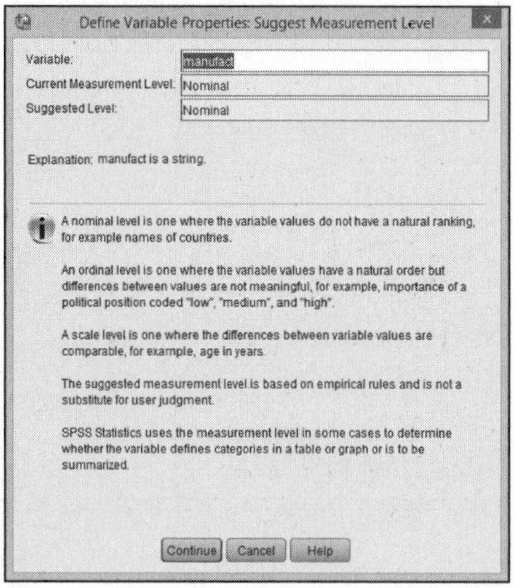

Figure 4-15:
From the
pattern
of values,
SPSS con-
cludes
whether you
may have
chosen the
wrong mea-
surement.

7. To apply any changes, click Continue.

You return to the window shown in Figure 4-14, where you can select another variable.

You won't want to make changes to all your variables, but SPSS helps you find the ones that you do need to change. Values defined as "missing" are not included in the computations. The text in the window always explains the criteria used to reach a conclusion, and SPSS allows you to make the final decision.

Chapter 5

Opening Data Files

• •

In This Chapter

▶ Understanding the SPSS file format

▶ Reading a simple text file into SPSS

▶ Transferring data from another application into SPSS

▶ Saving SPSS data in various formats

• •

*Y*ou don't need to put your data into the computer more than once. If you've entered your data in another program, you can copy it from there into SPSS — because every program worth using has some form of output that can serve as input to SPSS. This chapter discusses ways to transfer data into and out of SPSS.

Getting Acquainted with the SPSS File Format

SPSS has its own format for storing data and writes files with the .sav extension. This file format contains special codes and usually can't be used to export your data to another application. It's used only for saving SPSS data that you want to read back into SPSS at a later time. Several example files in this format are copied to your computer as part of the normal SPSS installation. These files can be found in the same directory as your SPSS installation. You can load any one of them by choosing File⇨Open⇨Data and selecting the file to be loaded. When you do so, the variable names and data are loaded and fill your SPSS window.

If you have SPSS filled with data, you can save it to a .sav file by choosing File⇨Save As and providing a name for the file. Or if you've loaded the information from a file, or you've previously saved a copy of the information to a file, you can simply choose File⇨Save to overwrite the previous file with a fresh copy of both variable definitions and data.

It's easy to be fooled by the way the SPSS documentation uses the word *file*. If you have defined data and variables in your program, the SPSS documentation often refers to it all as a "file," even though it may have never been written to disk. SPSS also refers to the material written to disk as a file, so watch the context.

When you write your file to disk, if you don't add the .sav extension to the filename, SPSS adds it for you. When you choose File➪Open➪Data to display the list of files, you may or may not see the extension on the filename (it depends on how your Windows system is configured), but it's there.

Formatting a Text File for Input into SPSS

If your data is in an application that can't directly create a file of a type that SPSS can read, getting the data into SPSS may be easier than you think. If you can get the information out of your application and into a text file, it's fairly easy to have SPSS read the text file.

When it comes to writing information to disk, some applications are more obliging than others. Look for an Export menu option — it usually has some options that allow you to organize the output text in a form you want. (Read on for a description of possible organization schemes.)

If the application doesn't allow you to format text the way you want, look for printer options — maybe you can redirect printer output to a disk file and work from there. If you use the application's printer output, you may need to use your word processor to clean up the form of the data. We know this multistep operation sounds like a lot of work, but it's often easier than typing all your data in again by hand.

The data file you output from SPSS doesn't have to include the variable names, just the values that go into the variables. You can format the data in the file by using spaces, tabs, commas, or semicolons to separate data items. Such dividers are known as *delimiters*. Another method of formatting data avoids delimiters altogether. In that method, you don't have to separate the individual data items, but you must make each data item a specific length, because you have to tell SPSS exactly how long each one is.

The most intuitive format is to have one case (one row of data) per line of text. That means the data items in your text file are in the same positions they'll be in when they're read into SPSS. Alternatively, you can have all your data formatted as one long stream, but you'll have to tell SPSS how many items go into each case.

Always save this kind of raw data as simple text; the file you store it in should have the .txt extension so SPSS can recognize it for what it is.

Reading Simple Data from a Text File

This section contains an example of a procedure you can follow to read data from a simple text file into SPSS. The file is named awards.txt. It contains two cases (rows of data) as two lines of text, with the data items in the two lines separated by spaces. The content of the file is as follows:

```
"Pat" 1 35 3.00 9
"Chris" 1 22 2.4 7
```

The following example reads this text file and inserts it into the cells of SPSS. Along the way, SPSS keeps you informed about what's going on so there won't be any big surprises at the end.

1. **Choose File⇨Read Text Data.**

   The Open Data window, shown in Figure 5-1, appears.

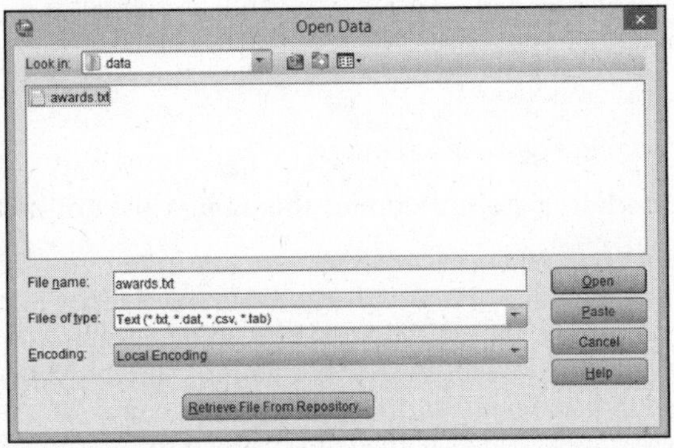

Figure 5-1: Locate the file you want to read.

2. **Select the awards.txt file, and then click Open.**

   The Text Import Wizard (the first screen of which is shown in Figure 5-2) appears, allowing you to load and format your data.

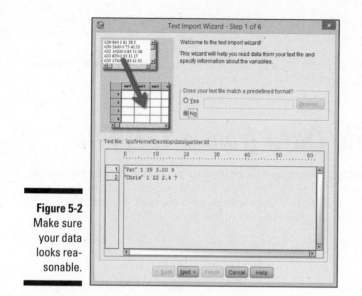

Figure 5-2
Make sure
your data
looks rea-
sonable.

3. Examine the input data.

The screen lets you peek at the contents of the input file so you can verify that you've chosen the right file. Also, if your file uses a predefined format (which it doesn't, in this example), you can select it here and skip some of the later steps. If your data doesn't show up nicely separated into values the way you want, you may be able to correct it in a later step. Don't panic just yet.

4. Click Next.

The screen shown in Figure 5-3 appears.

5. Specify that the data is delimited and the names are not included.

As you can see in this example, SPSS takes a guess, but you can also specify how your data is organized. It can be divided using spaces (as in this example), commas, tabs, semicolons, or some combination. Or your data may not be divided — it may be that all the data items are jammed together and each has a fixed width. If your text file includes the names of the variables (we show you how this works in a minute), you need to tell SPSS.

6. Click Next.

The screen shown in Figure 5-4 appears.

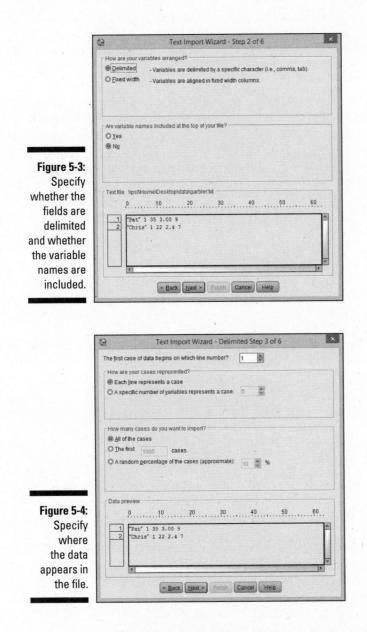

Figure 5-3:
Specify
whether the
fields are
delimited
and whether
the variable
names are
included.

Figure 5-4:
Specify
where
the data
appears in
the file.

7. Specify how SPSS is to interpret the text.

For this example, the correct settings are shown in Figure 5-4. You can tell SPSS something about the file and which data you want to read.

Perhaps some lines at the top of the file should be ignored — this happens when you're reading data from text intended for printing and header information is at the top. By telling SPSS about it, those first lines can be skipped.

Also, you can have one line of text represent one case (one row of data in SPSS), or you can have SPSS count the variables to determine where each row starts.

And you don't have to read the entire file — you can select a maximum number of lines to read starting at the beginning of the file, or you can select a percentage of the total and have lines of text randomly selected throughout the file. Specifying a limited selection can be useful if you have a large file and would like to test parts of it.

8. Click Next.

The screen shown in Figure 5-5 appears.

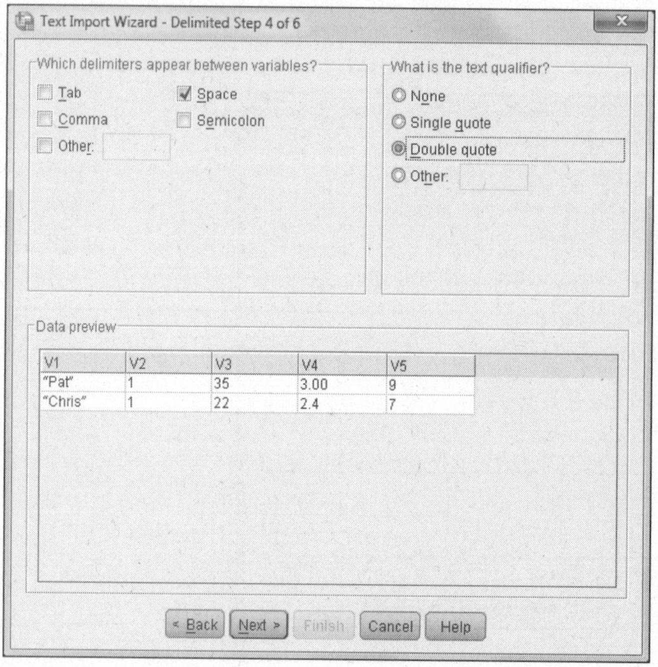

Figure 5-5: Specify the delimiters that go between data items and which quotes to use for strings.

9. Specify space as the delimiters and double quotes as text qualifiers.

SPSS knows how to use commas, spaces, tabs, and semicolons as delimiting characters. You can even use some other character as a delimiter by selecting Other and then typing the character into the blank. You can also specify whether your text is formatted with quotes (as in our example) and whether you use single or double quotes. Strings must be surrounded in quotes if they contain any of the characters being used as delimiters.

You can specify that a data item is missing in your text file. Simply use two delimiters in a row, without intervening data.

10. **Click Next.**

The screen shown in Figure 5-6 appears.

11. **Change the variable name and data format (optional).**

SPSS assigns the variables the names V1, V2, V3, and so on. To change a name, select it in the column heading at the bottom of the window, and then type the new name in the Variable Name field at the top. You can select the format from the Data Format drop-down list, as shown in Figure 5-6.

This step is optional. If you need to refine your data types, you can do so later in the Variable View tab of the Data Editor window. The point here is to get the data into SPSS.

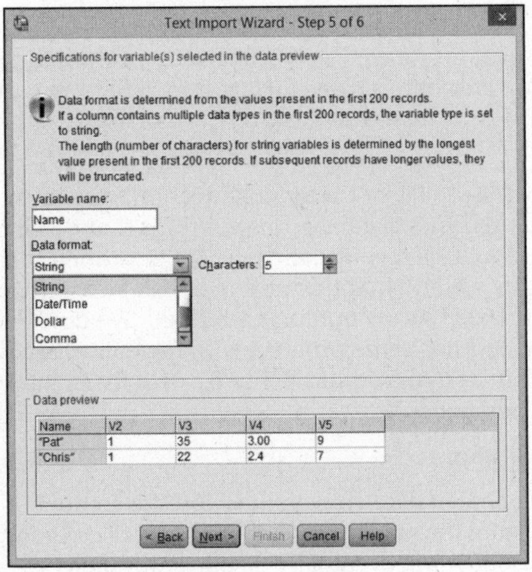

Figure 5-6:
Name your
variables
and select
their data
types.

12. **Click Next.**

The screen shown in Figure 5-7 appears.

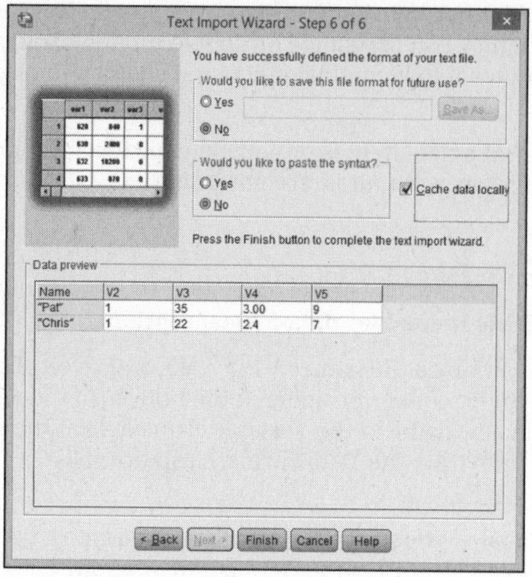

Figure 5-7:
Save the
format, grab
the syntax,
or enable
caching.

13. **In the Would You Like to Save This File Format for Future Use? Section, click No.**

    Saving the file format for future use is something you would do if you were loading more files of this same format into SPSS — it reduces the number of questions to answer and the amount of formatting to do next time.

    In the Would You Like to Paste the Syntax? section, you have the chance to grab a copy of the Syntax language instructions that do all this, but unless you know about the Syntax language (as described in Chapters 20 and 21), it's best to pretend that this option doesn't exist. (For that matter, the Cache Data Locally option is a bit odd. We don't know why it's there, unless SPSS has some problem with huge files. SPSS seems to load data faster with it than without it, but it's strictly an internal thing and SPSS works just fine either way.)

14. **Click the Finish button.**

    Depending on the type of data conversions and the amount of formatting, SPSS may take a bit of time to finish. But be patient. The Data View tab of the Data Editor window will eventually display your data.

15. **Look at the data. Correct your data types and formats, if necessary. Then save it all to a file by choosing File⇨Save As.**

    You're instructed to enter a filename. You can just call it Awards. The new file will have the .sav extension, which indicates that it's a standard SPSS file.

The SPSS way of reading data is a lot more flexible than this simple example demonstrates. Another example can help show why. Here, a file named `AwardHeader.txt` includes the same data, formatted slightly differently:

```
Name Sex Age GradePoint Awards
Pat,1,35,3.00,9,Chris,1,22,2.4,7
```

This time the data in the file is preceded by the variable names listed on the first line, the data is all in one long line, and the data is separated by commas. To read this into SPSS, you start the same way you did before. However, SPSS can't figure it all out in Step 1 this time (as shown in Figure 5-8). SPSS can't even tell which is header and which is data.

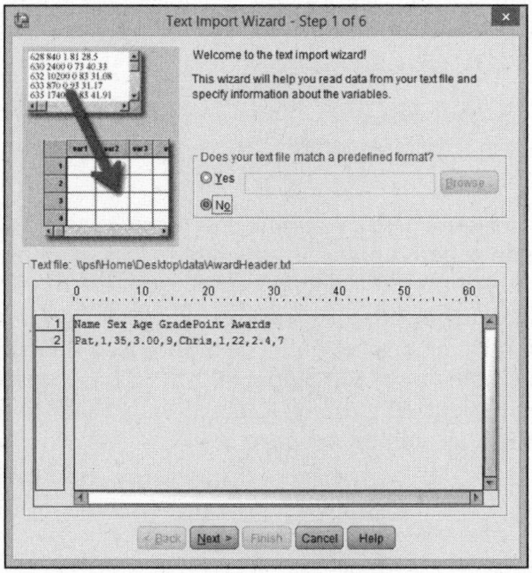

Figure 5-8:
The data remains as a block of text until you explain the parts.

In Step 2 of 6, you select the option that informs SPSS that the variable names appear in the first line of text. Then, in Step 3 of 6 (as shown in Figure 5-9), you specify that the data begins on line 2 of the text file and each case has five data items.

TIP

It's possible for the data to begin several lines down in the input text file, but if variable names are present, they must be on the first line. Also, when you specify variable names, SPSS ignores the beginning and ending of lines, and counts the data values to determine when it has a complete row (case).

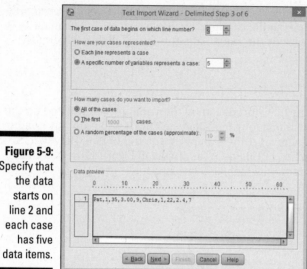

Figure 5-9:
Specify that
the data
starts on
line 2 and
each case
has five
data items.

In Step 4 of 6 (shown in Figure 5-10), commas and spaces were chosen as delimiters. (Although no spaces appear in the data in this example, it doesn't hurt to include a space delimiter if it may occur somewhere in your data.) Also, None was chosen for the characters surrounding string values. In this example, SPSS figured out the spacing on its own and used these settings for its default. Also, by the time you reach Step 4 of 6, SPSS has started organizing the data according to your definitions. It has already read the variable names and included them as column headers.

Figure 5-10:
Specifying
delimiters
and quote
characters.

In Step 5 of 6 (shown in Figure 5-11), you have the opportunity to change the variable names and specify their types. Here again, you see that SPSS has made a guess for the type of each one.

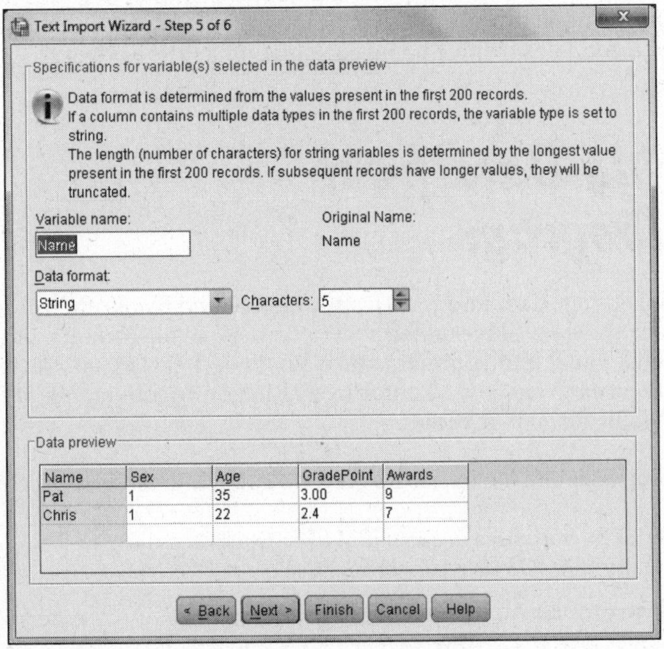

Figure 5-11: Specifications for variables.

After you complete Step 6 of 6, click the Finish button and wait for the data to load, as shown in Figure 5-12.

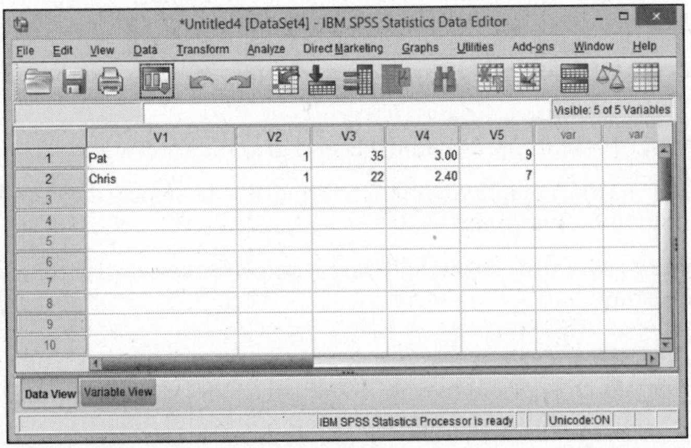

Figure 5-12: The data as formatted in SPSS.

You can see how many awards each person has, but you still have a little work to do. For example, click the Variable View tab, change the sex variable to a nominal data type, and assign the names "male" and "female" to the values 1 and 2. (You can't assume anything about sex by the names.) You may want to add some descriptive labels. For example, the awards variable could be given the descriptive name "number of awards won during lifetime." See how a good descriptive name can clear up a little mystery?

Transferring Data from Another Program

You can get your data into SPSS from a file created by another program, but it isn't always easy. SPSS knows how to read some file formats, but if you're not careful, you'll find your data stored in an odd file format. Deciphering some file formats can be as confusing as Klingon trigonometry. SPSS can read only from file formats it knows.

SPSS recognizes the file formats of several applications. Here's a complete list:

- **IBM SPSS Statistics (.sav):** IBM SPSS Statistics data, and also the format used by the DOS program SPSS/PC+

- **dBase (.dbf):** An interactive database system

- **Microsoft Excel (.xls):** A spreadsheet for performing calculations on numbers in a grid

- **Portable (.por):** A portable format read and written by other versions of SPSS, including other operating systems

- **Lotus (.w):** A spreadsheet for performing calculations with numbers in a grid

- **SAS (.sas7bdat, .sdy, .sd2, .ssd, and .xpt):** Statistical analysis software

- **Stata (.dta):** Statistical analysis and graphics software

- **Sylk (.slk):** A symbolic link file format for transporting data from one application to another

- **Systat (.syd and .sys):** Software that produces statistical and graphical results

Although SPSS knows how to read any of these formats, you may still need to make a decision from time to time about how SPSS should import your dataset. But you have some advantages:

✔ You know exactly what you want — the form of data appearing in SPSS is simple, and what you see is what you get.

✔ SPSS has some reasonable defaults and makes some good guesses along the way.

✔ You can always fiddle with things after you've loaded them.

You're only reading from the data file, so you can't hurt it. Besides, you have everything safely backed up, don't you? Just go for it. If the process gets hopelessly balled up, you can always call it quits and start over. That's the way we do it — we think of it as a learning process.

Reading an Excel file

SPSS knows how to read Excel files directly. If you want to read the data from an Excel file, we suggest you read the steps in "Reading Simple Data from a Text File," earlier in this chapter, because the two processes are similar. If you understand the decisions you have to make in reading a text file, reading from an Excel file will be duck soup. Figure 5-13 shows the appearance of data displayed by Excel.

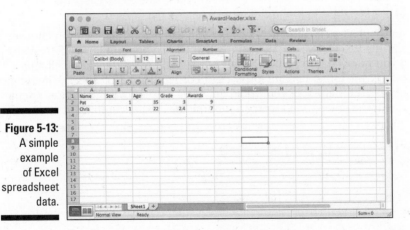

Figure 5-13:
A simple
example
of Excel
spreadsheet
data.

Do the following to read this data into SPSS:

1. Save the Excel data to a file.

In this example, the file is called `AwardHeader.xlsx`. If you want to copy only a portion of the spreadsheet, make a note of the cell numbers in the upper-left and lower-right corners of the group you want.

2. Close Excel.

You must stop the Excel program from running before you can access the file from SPSS.

3. Choose File⇨Open⇨Data.

4. Select the `.xlsx` file type, as shown in Figure 5-14, and then click Open.

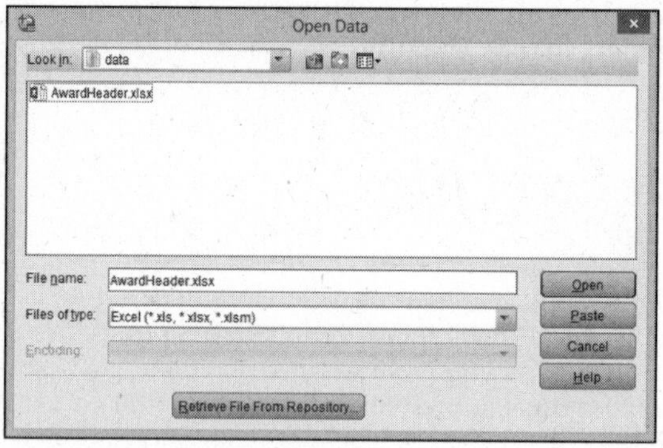

Figure 5-14:
From the many types of files understood by SPSS, select the Excel spreadsheet type.

5. Select the data to include.

An Excel file can contain more than one worksheet, and you can choose the one you want from the drop-down list, as shown in Figure 5-15. Also, if you've elected to read only part of the data, enter the Excel cell numbers of the upper-left and lower-right corners here. You specify the range of cells the same way you would in Excel — using two cell numbers separated by a colon. Don't worry about the maximum length for strings.

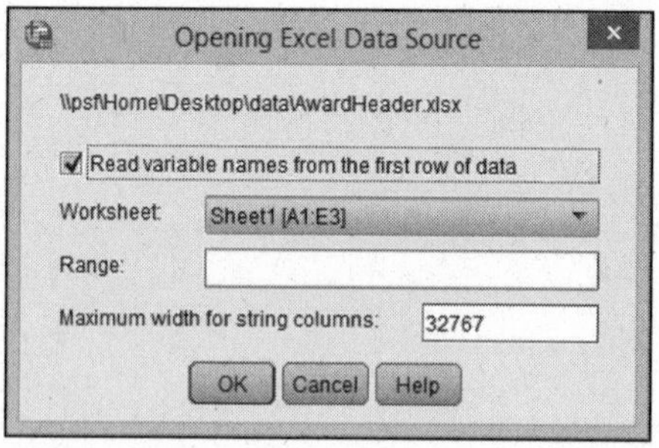

Figure 5-15:
Select which data in the spreadsheet to include.

6. Click OK.

Your data appears in the SPSS window.

7. **Check your variables and adjust their definitions as necessary.**

   SPSS makes a bunch of assumptions about your data, and it probably makes some wrong ones. Closely examine and adjust your variable definitions by switching to the Variable View tab and making the necessary changes.

8. **Save the file using your chosen SPSS name, and you're off and running.**

Reading from an unknown program type

Often, you can transfer data from another application into SPSS by selecting, copying, and pasting the data you want, but that method has its drawbacks. The places you're copying from and pasting to are usually larger than the screen, so highlighting and selecting can be tricky. You must be ready to choose Edit⇨Undo when necessary.

A better method is to write the data to a file in a format understood by SPSS, and then read that file into SPSS. SPSS knows how to read some file formats directly. Using such a file as an intermediary means you have an extra backup copy of your data, and that's never a bad idea.

Saving Data and Images

Writing data from SPSS is easier than reading data into SPSS. All you do is choose File⇨Save As, select your file type, and then enter a filename. You have lots of file types to choose from. You can write your data not only in two plain-text formats, but also in Excel spreadsheet format, three Lotus formats, three dBase formats, six SAS formats, and six Stata formats.

If you'll be exporting data from SPSS into another application, find out what kinds of files the other application can read, and then use SPSS to write in one of those formats.

A second form of output from SPSS is an image. If you've generated a graphic that you want to insert into your word processor or place on your website, SPSS is ready to help you do it. (We almost wish it were hard to do so we could look smart showing you how, but it's easy.)

When you go through the steps to produce a graph, as explained in Part IV, you'll be looking at the resulting graphics in the SPSS Statistics Viewer window, which is shown in Figure 5-16.

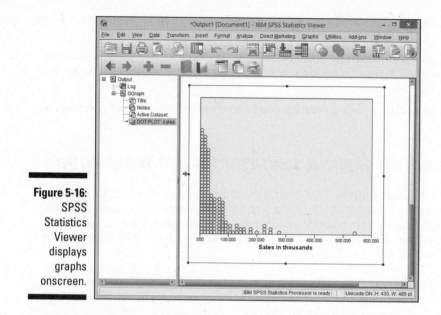

Figure 5-16:
SPSS
Statistics
Viewer
displays
graphs
onscreen.

From SPSS Statistics Viewer, you can export images (and do some other things, too). Here's how:

1. **Produce a graph or table.**

   You can use any of the examples in Part IV to produce a graphic display. The SPSS Statistics Viewer window pops up and displays the output.

2. **Choose File⇨Export.**

   The Export Output window, shown in Figure 5-17, appears.

3. **In the Objects to Export section, select which items to include in the output.**

   You can elect to have all objects output, all visible objects output, or only the ones you've selected. In Figure 5-16, for example, the panel on the left indicates that the graph is selected (because it's highlighted). The *visibility* of an object refers to whether its name appears in the list — if you collapse the list so a particular name can't be seen, the item is not visible. You can select items by clicking the items themselves, or by selecting their names in the list on the left.

4. **In the Document section, from the Type drop-down list, choose an output format.**

   Your choices vary according to what you decided to output as specified at the top of the window. Here's a list of the possible options:

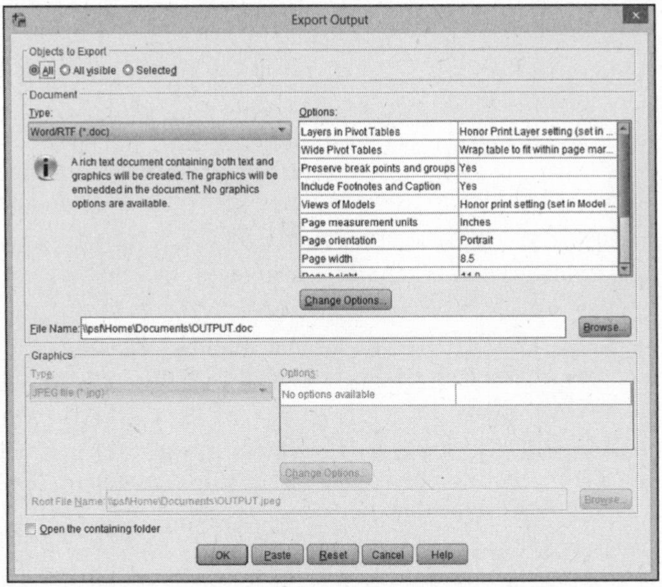

Figure 5-17:
These
selections
control what
gets output
and into
what format.

- *Excel 97–2004 (*.xls):* Excel files can include text, tables, and graphics, with the graphics embedded in the 97–2004 workbook. The data can create a new file or be added to an existing workbook. No graphic options are available.

- *Excel 2007 and higher (*.xlsx):* Excel files can include text, tables, and graphics, with the graphics embedded in the 2007 and higher workbook. The data can create a new file or be added to an existing workbook. No graphic options are available.

- *Excel 2007 and higher macro enabled (*.xlsm):* Excel files can include text, tables, and graphics, with the graphics embedded in the 2007 and higher macro-enabled workbook. The data can create a new file or be added to an existing workbook. No graphic options are available.

- *HTML (*.htm):* HTML files can be used for text both with and without graphics. If graphics are included, those will be exported separately, and they'll be included as HTML links. The graphic file type must also be chosen.

- *Web Reports (*.htm or *.mht):* Creates an interactive document that is compatible with most browsers, including Cognos Active Report.

- *Portable Document Format (*.pdf):* PDF documents exported will include not only text but also any graphics existing in the original. No graphics options are available.

- *PowerPoint (*.ppt):* PowerPoint documents can be written as text with the graphics embedded in the TIFF format. No graphic options are available.

- *Text-Plain (*.txt):* Text files can be output with graphic references included, and the graphics written to separate files. The reference is the name of the graphic file. The graphic file format is specified by choosing options in the lower section of this window.

- *Text-UTF8 (*.txt):* UTF-8 is Unicode text encoded as a stream of 8-bit characters. Graphics are handled the same as they are for text files.

- *Text-UTF16 (*.txt):* UTF-16 is Unicode text encoded as a stream of 16-bit characters. Graphics are handled the same as they are for text files.

- *Word/RTF (*.doc):* Word documents are written in rich text format (RTF), which can be copied into a Word document. No graphic options are available.

- *None:* When selected, this option means no text is output — only graphic images. The graphic file format is specified by options in the lower section of this window.

5. **In the Graphics section, select the image file format, if one is needed, from the Type drop-down list.**

   You may be asked to select a format for your image file(s). You can select from .png, .bmp, .emf, .eps, .jpg, or .tif.

6. **Click the Browse button, select the directory and root filename, and click Save.**

   Depending on what you chose to output, the actual output may be multiple files, and they'll all have names derived from the root name you provide. The Save button doesn't write the file(s) — it only inserts your selected name into the Export Output window.

7. **Click OK.**

   The file(s) are written to disk — each in the chosen format, at the chosen location.

Chapter 6

Getting Data and Results out of SPSS

SPSS is good at analyzing your data and displaying information that's easy to understand in tables, charts, and graphs. But the time comes when you want to output the results to files suitable for use in other applications. You may want to send output to the printer, or you may have another program that could make use of the output from SPSS. This chapter explains ways that you can get your data out of SPSS and into the forms that other programs need.

Printing

The simplest form of output is to print the numeric rows and columns of the raw data as it appears on the Data View tab of the Data Editor window. To do so, choose File➪Print. A familiar Print dialog box appears, where you can select the print settings you need for your system. The table of data will be printed with lines between the rows and columns, the same as they appear onscreen. The printed form has case numbers to the left and variable names at the top.

If you're not sure what your output will look like, you can choose File➪Print Preview and see, on the screen, the same layout that will be sent to the printer. The zoom and page-selection controls at the top of the window allow you to examine the output.

If the table you're printing is too wide to fit on the sheet of paper, SPSS splits the output and places the table on multiple pages. You can hold the printed sheets side by side to get the full width of the table.

If you want to print the variable definitions, you can switch from the Data View tab to the Variable View tab before printing. This output always requires two pages because it includes the full width of the table.

Exporting to a Database

You can export SPSS data directly to a database. Choose File⇨Export⇨Export to Database and follow the instructions SPSS supplies for your database. SPSS knows how to write to Access, dBase, Excel, FoxPro, and text file databases. If you have a different database system, you should be able to configure SPSS for it by clicking the Add ODBS Data Source button. You should be able to get the information you need to do this from the documentation of your database. In similar fashion, you can read data from a database by choosing File⇨Open Database.

To export the data, simply follow the onscreen instructions for selecting the variables to be written and for choosing whether to append new data or to overwrite existing data.

Using SPSS Statistics Viewer

Whenever you run an analysis, produce a graph, or do anything that generates output (even loading a file), the SPSS Statistics Viewer window pops up automatically to display what you've created. This display is the most fundamental form of output from SPSS and is the first step in producing other forms of output.

Chapters 11, 12, 18, and 19 provide details about generating tables, graphs, and descriptive text in SPSS Statistics Viewer. These chapters describe how to output Viewer data to files in different formats.

You can output data from SPSS Statistics Viewer in several file formats appropriate for use by other applications. Some output formats are graphics only, some are text only, and others are a mixture of text and graphics. Some form of graphic output is usually necessary because of the graphs and charts constructed by SPSS.

In every case, you begin by choosing File⇨Export from the menu of SPSS Statistics Viewer, which displays the Export Output dialog box (shown in Figure 6-1). In the Export drop-down list, you can choose which items in the View window to export — the entire document, the text of the document without graphics, or the graphics without text.

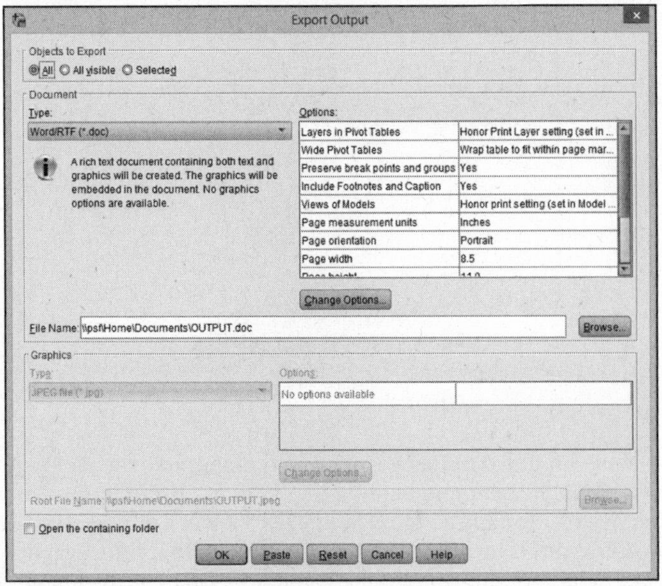

Figure 6-1:
The main control window for generating output from SPSS Statistics Viewer.

> **Export Output**
>
> Objects to Export
> ⦿ All ○ All visible ○ Selected
>
> Document
> Type:
> Word/RTF (*.doc)
>
> ⓘ A rich text document containing both text and graphics will be created. The graphics will be embedded in the document. No graphics options are available.
>
> Options:
>
> | | |
> |---|---|
> | Layers in Pivot Tables | Honor Print Layer setting (set in ... |
> | Wide Pivot Tables | Wrap table to fit within page mar... |
> | Preserve break points and groups | Yes |
> | Include Footnotes and Caption | Yes |
> | Views of Models | Honor print setting (set in Model ... |
> | Page measurement units | Inches |
> | Page orientation | Portrait |
> | Page width | 8.5 |
> | Page height | 11.0 |
>
> Change Options...
>
> File Name: \\psf\Home\Documents\OUTPUT.doc   Browse...
>
> Graphics
> Type:
> JPEG file (*.jpg)
> Options:
> No options available
>
> Change Options...
>
> Root File Name: \\psf\Home\Documents\OUTPUT.jpeg   Browse...
>
> ☐ Open the containing folder
>
> OK   Paste   Reset   Cancel   Help

At the very top of the dialog box, you can select which pieces of information in SPSS Statistics Viewer you want to include as part of the output:

- ✔ **All:** Outputs all the information that SPSS Statistics Viewer contains, regardless of whether the information is currently visible

- ✔ **All Visible:** Includes only those objects being displayed by SPSS Statistics Viewer

- ✔ **Selected:** Allows you to select which objects to output

The set of selections made available to you in the Export Output dialog box is determined by the types of objects being displayed by SPSS Statistics Viewer, which (if any) are selected, and the choice in the Export drop-down list. The only combinations of options available are those that produce output.

Figure 6-2 shows an SPSS Statistics Viewer window displaying both text and graphics. On the left is a list of names of objects. If the name of an object is visible in the list, the object itself is visible in the SPSS Statistics Viewer

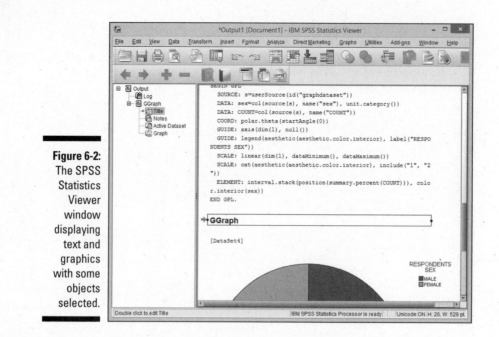

Figure 6-2:
The SPSS
Statistics
Viewer
window
displaying
text and
graphics
with some
objects
selected.

window. You can make objects appear and disappear by clicking the plus
and minus signs. If the name of an object is highlighted in the list, the object
is marked as selected in the SPSS Statistics Viewer window; a selected object
appears surrounded by boxes. (In the figure, the log, title, and notes at the
top are not selected, but the other objects are.) When producing output, you
can select to export only visible objects, only selected objects, or all objects.

You can output the following types of files:

- ✔ Plain text

- ✔ Unicode (UTF-8 or UTF-16)

- ✔ HTML web page

- ✔ Microsoft Excel file

- ✔ Rich text format (RTF), readable by Microsoft Word

- ✔ Microsoft PowerPoint display file

- ✔ Portable document format (PDF) file

Some formats (for example, the text-file format) require that graphics be
output in separate files; you can also elect to output *only* graphics files.
Graphics can be output in the following formats:

- ✔ Standard jpeg (JPG)

- ✔ Portable network graphics (PNG)

✔ Postscript (EPS)

✔ Tagged image file format (TIFF)

✔ Windows bitmap (BMP)

✔ Enhanced metafile (EMF)

Simple copy and paste

When you've run an analysis and produced a graph in SPSS Statistics Viewer, the simplest way to transfer the graphic to Microsoft Word or another document type is using Copy and Paste.

When you have output (refer to Figure 6-2), scroll to the graphic. To copy the graphic, right-click the image and click Copy, as shown in Figure 6-3. Open a new Microsoft Word document and paste the graphic. To do this, right-click in the document and click Paste. Depending on the settings on your computer, you can copy the graphic using Ctrl+C (⌘+C) and paste it using Ctrl+V (⌘+V).

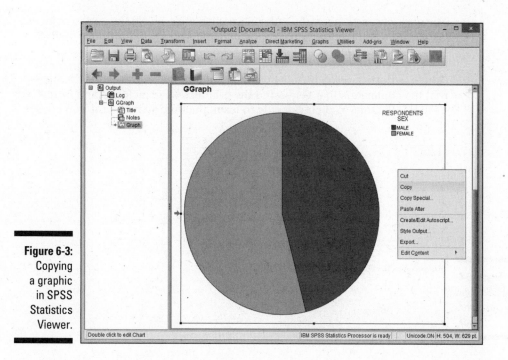

Figure 6-3:
Copying
a graphic
in SPSS
Statistics
Viewer.

Creating an HTML web page file

If you decide to format your output file as a web page, the output text will be formatted as HTML. Any pivot tables selected for output will be formatted as HTML tables, and any images to be output will be written to separate files in the image format of your choice. (A description of the image-file options appears at the end of this chapter.)

You can make a number of decisions about the details of the HTML file; they appear when you click the Change Options button for the document in the Export Output dialog box.

The first options to set are the layers in pivot tables. Some pivot tables have more than two dimensions, and the other dimensions are presented as multiple displayable layers. By setting this option, you can include or exclude layers in the HTML file — if you have a multilayered table, you'll probably have to experiment with this setting to get desirable results.

A pivot table can have multiple headings and footnotes. You can choose to have the footnotes and headings included or excluded.

The onscreen view is not the only one available. You have the option of including all views in the output or showing only the view that is currently visible.

Figure 6-4 shows part of the output page as it appears in a web browser — using the default settings for everything, including the JPEG image. Notice that the commands that generated the graphics were included and formatted in an HTML table. You may decide to leave that information out. You could,

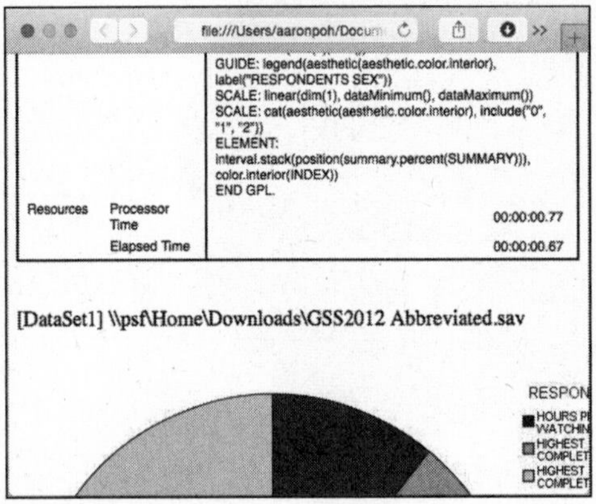

Figure 6-4: SPSS output as a web page.

if you wanted, leave the table out and include only the graphic and its annotation. Also, if you were going to publish this as a web page, you'd probably want to edit the heading so it was something other than the name of a working directory on your local machine.

In this example, the output filename is `GSS2012Abbreviated`, so the main file is `GSS2012Abbreviated.htm` and the image file is `GSS2012Abbreviated.jpg`. The `.jpg` suffix indicates that a JPEG image file was the chosen option. The digit in the image filename is necessary because there could be more than one; each needs a unique name.

Creating a text file

If you want to output a simple text file, you still have a number of options to choose from, as shown in Figure 6-5. The first two options are whether to use tabs or spaces to position characters on the page. This choice can be important because alignment is crucial to some data layouts, and programs that read the text files may have different tab settings and change the appearance of the output when it's displayed.

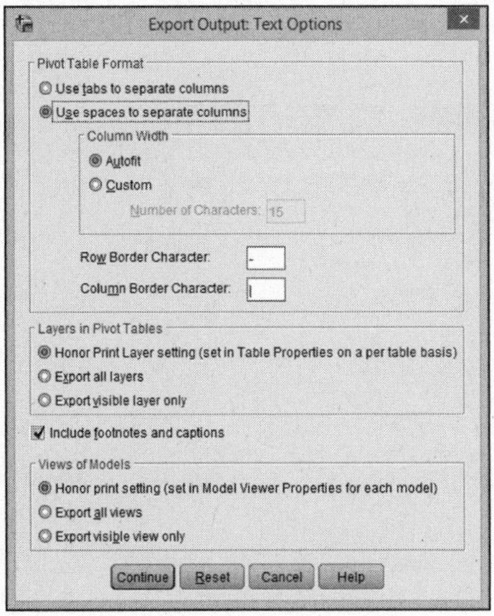

Figure 6-5:
The options
for creating
a text file.

The options for creating a Unicode file are the same as those for creating a plain-text file. The Unicode output is in the standard encoding format of your choice — either UTF-8 or UTF-16. You would want to output text in one of these formats only if you have a program that needs one of those formats for its input.

Tables output as text use certain characters to define the cells in which data items are shown. You can select any characters you want to act as separators and draw the borders, or you can accept the default of the minus sign and vertical bar, as shown in the figure. (The vertical bar is a standard keyboard character, usually on the same key as the backward slash. It sometimes looks like a vertical line broken in the middle.) If you're outputting tables, you can choose a maximum cell size or just use the default Autofit option and let SPSS decide the number of characters that will fit in each column.

Creating an Excel file

Creating an Excel file is easier than creating either a text file or an HTML file because the images are not generated as separate files — graphic images are included in the worksheet. (The options for creating an Excel file are shown in Figure 6-6.) You get to choose how pivot tables, footnotes, captions, and models are handled, as described for HTML files.

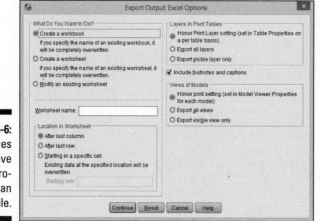

Figure 6-6:
The choices you have when producing an Excel file.

By default, a new workbook file is created. If a file with the same name already exists, it's overwritten. You can specify that a new worksheet be created within the workbook file; however you must specify the name of the worksheet. If the worksheet name you choose already exists in the workbook, the file that has it is overwritten. Alternatively, you can specify that the output be used to modify an existing worksheet within the workbook file. If you decide to specify the name of a worksheet, the name can't exceed 32 characters and shouldn't include any special characters (anything other than numbers and letters).

Also, if you choose to modify a worksheet, you can specify where, in the existing worksheet, the new information is to be placed.

When you want to produce output, click OK in the Export Output dialog box, and a file is generated. Then you can load the file directly into Excel, as shown in Figure 6-7.

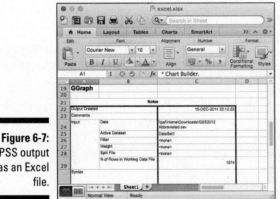

Figure 6-7: SPSS output as an Excel file.

In this example, the output filename is excel, so the output file was named excel.xlsx.

Creating a Word document file

If you choose to output a Word document file, you have no graphic options to set because both text and graphics are included in one output file. The options you can choose from are shown in Figure 6-8: whether to include all layers of any tables that may be in the output, whether to include footnotes and captions, and how models are to be handled.

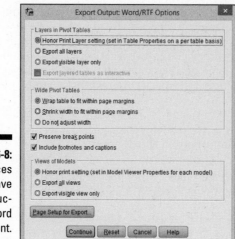

Figure 6-8:
The choices
you have
in produc-
ing a Word
document.

The Page Setup for Export button opens a dialog box that allows you to lay out the page size and margins of the output. It makes it possible to specify wrapping and shrinking to make things fit.

When you want to produce output, click Continue in the Export Output dialog box, and the file is generated. Then you can load the output file directly into Word, as shown in Figure 6-9.

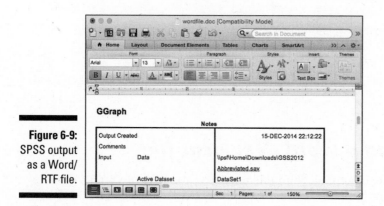

Figure 6-9:
SPSS output
as a Word/
RTF file.

In this example, the output filename is `wordfile`, so the output file was named `wordfile.doc`.

If the output file is in rich text format (RTF), it can be loaded and used by most word processors, including OpenOffice, StarOffice, and WordPerfect.

Creating a PowerPoint slide document

A PowerPoint file includes only tables, graphs, and models, so you can produce a series of display slides that contain all your graphics. The basic options are shown in Figure 6-10.

Export Output: PowerPoint Options

Layers in Pivot Tables
- ● Honor Print Layer setting (set in Table Properties on a per table basis)
- ○ Export all layers
- ○ Export visible layer only
- ☐ Export layered tables as interactive

Wide Pivot Tables
- ● Wrap table to fit within page margins
- ○ Shrink width to fit within page margins
- ○ Do not adjust width

☑ Include footnotes and captions
☐ Use Viewer outline entries as slide titles

Views of Models
- ● Honor print setting (set in Model Viewer Properties for each model)
- ○ Export all views
- ○ Export visible view only

Page Setup for Export...

Continue   Reset   Cancel   Help

The first options to set are the layers in pivot tables. Some pivot tables have more than two dimensions, and the other dimensions are presented as multiple display layers. By setting this option, you can include or exclude layers in the PowerPoint slides. If you have a multilayered table, you'll probably have to experiment with the setting and see what you get.

A pivot table can have multiple headings and footnotes, which you can choose to include or exclude. You can also choose to use the outline headings as slide titles in your output.

Some results are presented as graphic models, but the view that appears onscreen isn't the only one you can use. You have the option of including all views in the output, or showing only the view that is currently visible.

Your output will include only charts, graphs, and pivot tables; the rest of your data is ignored and doesn't appear anywhere in the set of produced slides. Figure 6-11 displays the slide produced from the same SPSS Statistics Viewer data that was used in the previous example. If you need some text slides before or after your graphics, you have to add those yourself.

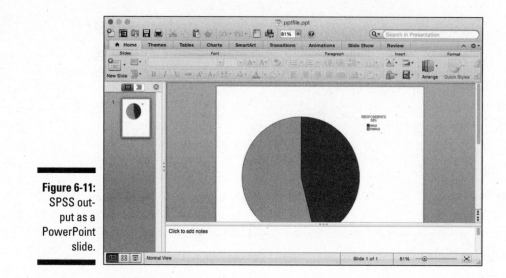

In this example, the output filename is `pptfile`, so the output file was named `pptfile.ppt`.

Creating a PDF document

It's becoming more common to place information on the Internet in a PDF format instead of an HTML format. Both are read-only files, but a PDF gives the creator of the file more control over the document's appearance when it's displayed in a viewer. An HTML page is relatively freeform compared to a PDF file. With a PDF file, you can put your information on the Internet and have it seen the same way by every person who views it.

A PDF file contains both formatted text and graphics, so any PDF you output will look very much like the original data displayed in SPSS Statistics Viewer. PDF handles graphics in a standard way, so you don't have the typical graphic options to set. Note, however, that you do have some other options, as shown in Figure 6-12.

You can elect to include bookmarks in the produced file. These bookmarks are important for larger files. They're used by the viewer to simplify the process of navigating through the file.

Embedding fonts ensures that the document will look the same on every computer. If the fonts aren't embedded, the chosen font may not be available for display or print, in which case the substitute font could make the resulting display look quite different.

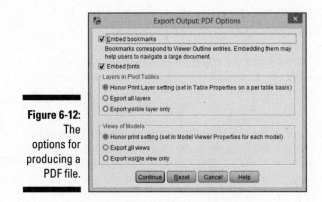

Figure 6-12:
The
options for
producing a
PDF file.

You can set the layers in pivot tables. Some pivot tables have more than two dimensions, and the other dimensions are presented as multiple displayable layers. By setting this option, you can include or exclude layers in the PDF file. If you have a multilayered table, experiment with this setting until you get the results you want.

Using the default settings, SPSS produced a PDF file, shown in Adobe Acrobat in Figure 6-13.

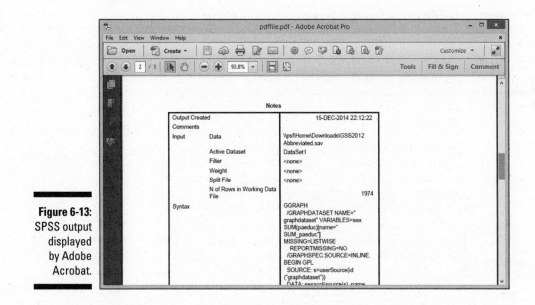

Figure 6-13:
SPSS output
displayed
by Adobe
Acrobat.

Creating a Graphics File

Depending on the type of output data file you generate, you may need to select the file type and configuration settings for separate image files. When you produce such image files, you could get several of them — one for each image displayed in SPSS Statistics Viewer. And image files have options.

If you don't have an immediate handle on the options you can use to generate your selected type of graphics file, experiment. Start with the defaults and make changes only if you need to. It doesn't cost anything to try different combinations of options and decide on the settings you like.

For all image file types, you can specify the size in terms of a percentage of the original. The default is 100%, which means there is no change in the size of the image. The other options available (compression, number of colors, and so on) vary, depending on the file type.

Figure 6-14 is the dialog box used to set the options for a bitmap (.bmp) file. The size can be expanded up to 200% of the size displayed in SPSS Statistics Viewer. You can also choose to use compression to reduce the size of the file — the compression used won't reduce the quality of the image.

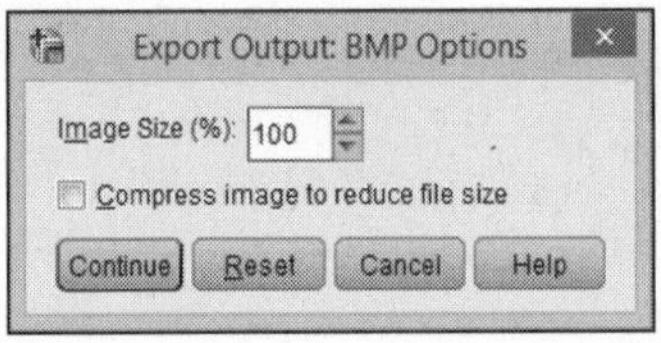

Figure 6-14:
Options for configuring BMP files.

> Export Output: BMP Options
>
> Image Size (%): 100
>
> ☐ Compress image to reduce file size
>
> Continue   Reset   Cancel   Help

Figure 6-15 is the dialog box used to set the options for an enhanced metafile (.emf). The only option is to adjust the size of the image.

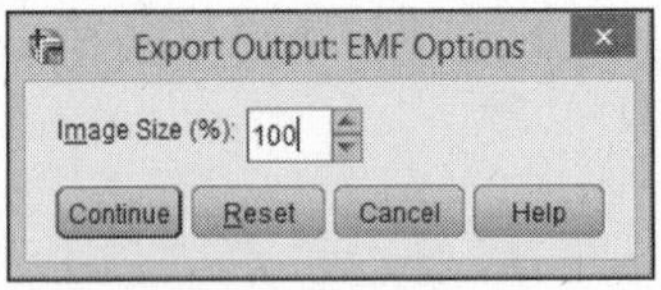

Figure 6-15:
Options for configuring EMF files.

> Export Output: EMF Options
>
> Image Size (%): 100
>
> Continue   Reset   Cancel   Help

Figure 6-16 is the dialog box used to set the options for an encapsulated postscript (.eps) file. You can set the size of the image in one of two ways — you can set it to be a percentage of the current size or you can set its width as a number of points. (There are 72 points to an inch.) You can optionally choose to produce a TIFF along with the postscript image, in case you're unable to display the postscript image. If the fonts are all available on the output device, you can simply include the font information. If the fonts aren't available, a substitute will be chosen. Alternatively, you can choose to present the fonts as a collection of graphics (curves).

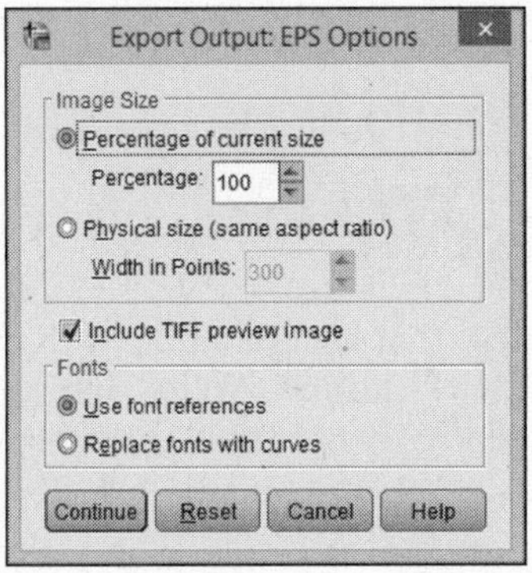

Figure 6-16:
Options for configuring EPS files.

Figure 6-17 is the dialog box used to set the options for a JPEG (.jpg) file. You can set the size, or you can choose to remove color from the image.

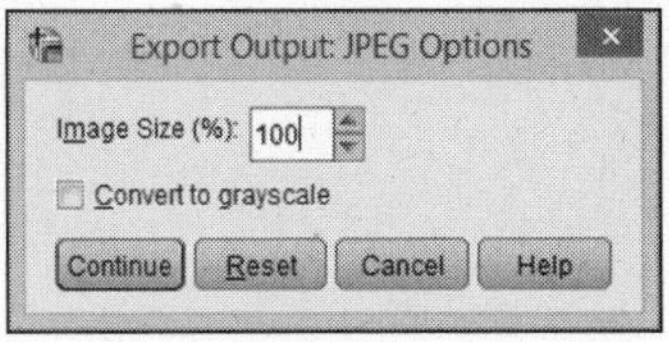

Figure 6-17:
Options for configuring JPEG files.

Figure 6-18 is the dialog box used to set the options for a PNG (.png) file. You can set the size as a percentage of the original. The color depth determines the maximum number of colors that can be used in the display. If the output is composed of fewer colors than appear in the original, the output is dithered to differentiate the graphics.

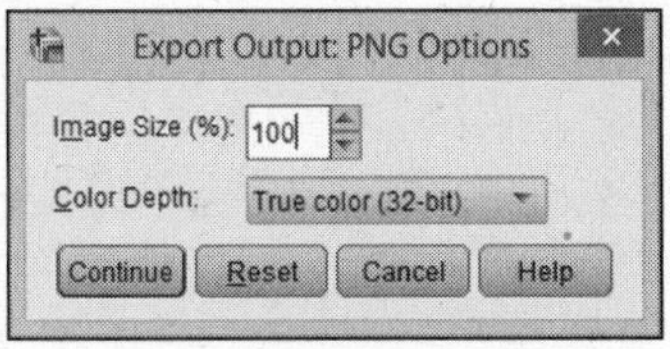

Figure 6-18:
Options for configuring PNG files.

Figure 6-19 is the dialog box used to set the options for a TIFF (.tif) file. The option you can set is the size as a percentage of the original.

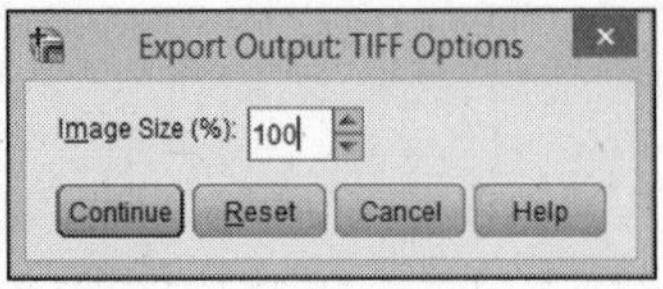

Figure 6-19:
Options for
configuring
TIFF files.

Figure 6-19:
Options for
configuring
TIFF files.

Creating a Web Report File

A web report is an interactive document that is compatible with most web browsers, including popular ones such as Chrome, Firefox, Internet Explorer, and Safari. Web reports allow the flexibility to see reports online and, unlike PDFs, are interactive.

A web report file contains both formatted text and graphics, so any web report you output will look very much like the original data displayed in SPSS Statistics Viewer. Web report files handle graphics in a standard way, so you don't have the typical graphic options to set. Note, however, that you do have some other options, as shown in Figure 6-20. Web reports come in two formats: SPSS Web Report (HTML5) and Cognos Active Report (.mht).

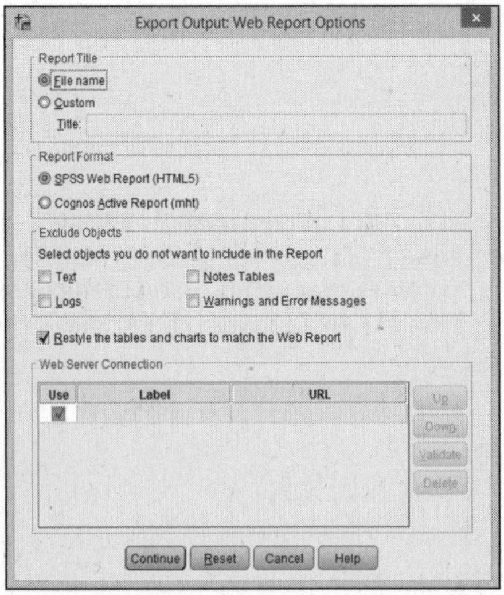

Figure 6-20:
Options for
producing a
web report.

Figure 6-21 shows the interactive HTML5 Web Report file using Internet Explorer. The interactive menu bar on the left side of Figure 6-21 allows you to click through the Log, Notes, and Graph portions of the web report.

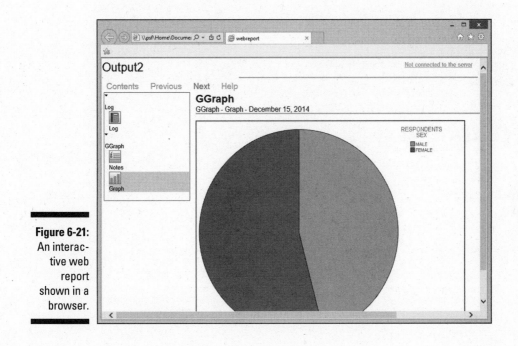

Figure 6-21: An interactive web report shown in a browser.

Chapter 7

More About Defining Your Data

*W*ithout a definition, a number serves no purpose. For example, the number 3 could have entirely different meanings. It could be a number of miles, or an answer to a multiple-choice question, or the number of jelly beans in your pocket.

The data type is more than just a tag — it determines how the value can be manipulated. For example, *date arithmetic* (the distance in time between two dates) would be a nightmare to do without help. You would have to take into account leap years, and you may even have to worry about whether a day is a business day. So, it isn't enough to merely declare that something is a date, as we do in Chapter 4. You have to make sure that the date is in the proper date format. As soon as you do that, you can take advantage of special menus for manipulating dates.

Multiple-response variables — those "check all that apply" questions on surveys — are another kind of variable type that needs extra attention. Again, when you do it properly, you can use a special menu dedicated to this type of variable. Finally, all this data definition stuff can be time consuming, so there is a special shortcut menu for copying your definition work from one dataset to another.

Working with Dates and Times

Calendar and clock arithmetic can be tricky, but SPSS can handle it all for you. Just enter the date and time in whatever format you specify, and SPSS converts those values into its internal form to do the calculations. Also, SPSS displays the date and time in your specified format, so it's easy to read.

SPSS understands the meaning of slashes, commas, colons, blanks, and names in the dates and times you enter, so you can write the date and time almost any way you'd like. If SPSS can't figure out what you've typed, it clears away what you typed and waits for you to type something again.

Internally, SPSS keeps all dates as a positive or negative count of the number of seconds from a zero date. Here's a bit of trivia for you. The zero date in SPSS is the birth of the Gregorian calendar in 1582. No kidding! You can choose a display format that includes or excludes the time, but the information is always there. You can even change the display format without loss of data. If the time isn't included in the data you enter, SPSS assumes zero hours and minutes (midnight).

You determine the data type for each variable on the Data View tab of the Data Editor window. The type is chosen from the list of types shown in Figure 7-1. On the right, you select a format. SPSS uses this format to interpret your input and to format the dates for display.

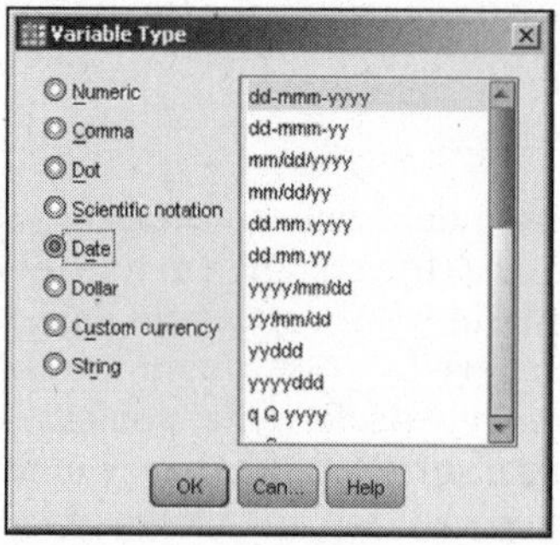

Figure 7-1:
Select the
data type
and the
format.

SPSS uses the format you select for both reading your input and formatting the output of dates and times.

The Columns setting of the date variable on the Variable View tab of the Data Editor window is important. The column width determines the maximum number of characters that can be displayed. If you choose a format that is too narrow to fit, the date will show up only as a row of asterisks.

The available formats are defined as a group and change according to the variable type. For example, the Dollar type has a different list of choices from those offered for the Date type.

The list of format definitions you have to choose from are constructed by combining the specifiers listed in Table 7-1. Format definitions look like mm/dd/yy and ddd:hh:mm.

Table 7-1	Specifiers in Date and Time Formats
Specifier	*Means*
dd	A two-digit day of the month in the range 01, 02, . . . , 30, 31.
ddd	A three-digit day of the year in the range 001, 002, . . . , 364, 365.
hh	A two-digit hour of the day in the range 00, 01, . . . , 22, 23.
Jan, Feb, . . .	The abbreviated name of the month of the year, as in JAN, FEB, . . . , NOV, DEC.
January, February, . . .	The name of the month of the year, as in JANUARY, FEBRUARY, . . . , NOVEMBER, DECEMBER.
mm	When adjacent to a dd specifier in a format, a two-digit month of the year in the range 01, 02, . . . , 11, 12. When adjacent to an hh specifier in a format, a two-digit specifier of the minute in the range 00, 01, . . . , 58, 59.
mmm	A three-character name of a month, as in JAN, FEB, . . . , NOV, DEC.
Mon, Tue, . . .	The abbreviated name of the day of the week, as in MON, TUE, . . . , SAT, SUN.
Monday, Tuesday, . . .	The name of the day of the week, as in MONDAY, TUESDAY, . . . , SATURDAY, SUNDAY.
q Q	The quarter of the year, as in 1 Q, 2 Q, 3 Q, or 4 Q.
Ss	Following a colon, the number of seconds in the range 00, 01, . . . , 58, 59. Following a period, the number of hundredths of a second.
ww WK	The one- or two-digit number of the week of the year in the range 1 WK, 2 WK, . . . , 51 WK, 52 WK. *Note:* Although week numbers can be either one or two digits, the numbers always line up when printed in columns because SPSS inserts a blank in front of single-digit numbers.
yy	A two-digit year in the range 00, 01, . . . , 98, 99. The assumed first two digits of the four-digit year this represents are determined by the configuration found at Edit⟹Options⟹Data.
yyyy	A four-digit year in the range 0001, 0002, . . . , 9998, 9999.

You can go back and change the format of a date variable at any time without fear of losing information. For example, you could enter the data under a format that accepted only the year, month, and day, and then change the format to something that contains only the hours and minutes. The format may not display all the information you entered (in fact, in this case, it won't), but when you change the format back to something more inclusive, all your data is still there.

To enter data, choose a format — any format — that contains all the data you have. You can later change to a more limited format that displays only the information you want. But you can't go the other way. If you later choose a format that doesn't leave parts out, you see the defaults that were inserted by SPSS when you entered the data.

Using the Date Time Wizard

If you have dates that have been properly declared, you can easily do numerous types of calculations. Just follow these steps:

1. **Open the `nenana2.sav` dataset.**

   This dataset is similar to the `nenana.sav` dataset except that a date time stamp has been created using the original variable, just like the demonstration in Chapter 3.

2. **Choose Transform⇨Date and Time Wizard.**

   The window shown in Figure 7-2 appears.

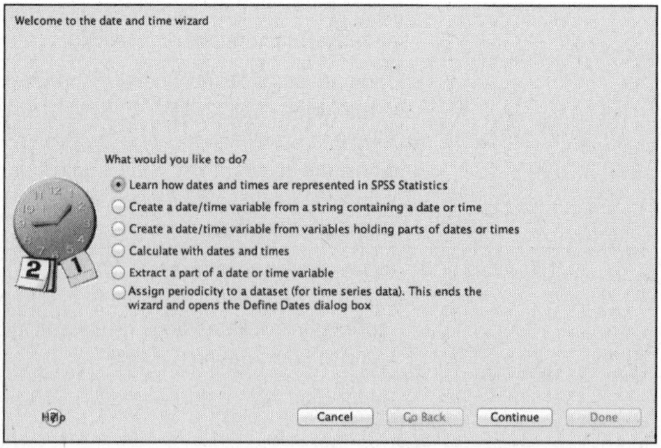

Welcome to the date and time wizard

What would you like to do?

- ● Learn how dates and times are represented in SPSS Statistics
- ○ Create a date/time variable from a string containing a date or time
- ○ Create a date/time variable from variables holding parts of dates or times
- ○ Calculate with dates and times
- ○ Extract a part of a date or time variable
- ○ Assign periodicity to a dataset (for time series data). This ends the wizard and opens the Define Dates dialog box

Help   Cancel   Go Back   Continue   Done

Figure 7-2:
The Date and Time Wizard.

3. **Select the Extract a Part of a Date or Time Variable radio button, and click Continue.**

4. **Choose DateTime as the Date or Time variable and Day of Week as the Unit to Extract (see Figure 7-3), and click Continue.**

5. **Call the Result Variable the new name Day_of_Week2, and click Done.**

    You can check your work against Figure 7-5, if you like, but we'll do a second calculation now. Figure 7-5 shows both calculations.

6. **Return to Transform⇨Date and Time Wizard.**

    The window shown in Figure 7-2 appears again.

7. **Select the Calculate with Dates and Times radio button this time, and click Continue.**

8. **Choose Current Date and Time [$TIME] as Date1 and DateTime as the Date2.**

9. **Select the Retain Fractional Part radio button, and click Continue.**

10. **Call the Result Variable the new name Years_Since2, and click Finish.**

    The selections are shown in Figure 7-4.

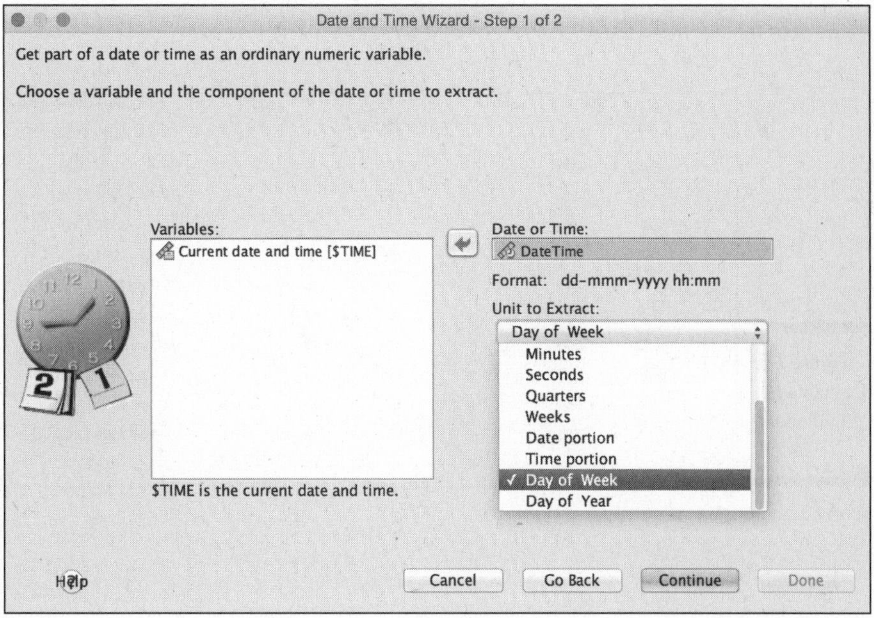

Figure 7-3: Extracting Day of Week.

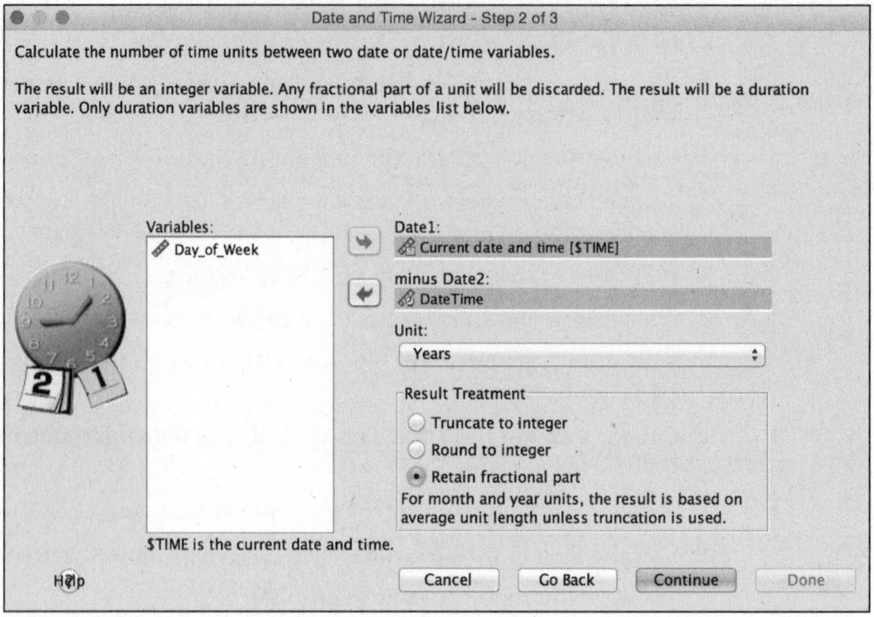

Figure 7-4:
Date
subtraction.

11. Check your work against Figure 7-5.

	year	month	day	hour	minute	DateTime	Day_of_Week	Years_Since
1	1940	4	20	15	27	20–Apr–1940 15:27	Sat	74.84
2	1998	4	20	16	54	20–Apr–1998 16:54	Mon	16.85
3	1993	4	23	13	1	23–Apr–1993 13:01	Fri	21.84
4	1990	4	24	17	19	24–Apr–1990 17:19	Tue	24.84
5	2004	4	24	14	16	24–Apr–2004 14:16	Sat	10.83
6	1926	4	26	16	3	26–Apr–1926 16:03	Mon	88.83
7	1995	4	26	13	22	26–Apr–1995 13:22	Wed	19.83
8	1988	4	27	9	15	27–Apr–1988 09:15	Wed	26.83
9	1943	4	28	19	22	28–Apr–1943 19:22	Wed	71.82
10	1969	4	28	12	28	28–Apr–1969 12:28	Mon	45.82
11	2005	4	28	12	1	28–Apr–2005 12:01	Thu	9.82
12	1939	4	29	13	26	29–Apr–1939 13:26	Sat	75.82

Figure 7-5:
Dataset with
calculations
added.

Creating and Using a Multiple-Response Set

A *multiple-response set* is much like a new variable made of other variables you already have. A multiple-response set acts like a variable in some ways, but in other ways it doesn't. You define it based on the variables you've already defined, but it doesn't show up on the Variable View tab. It doesn't even show up when you list your data on the Data View tab. But it does show up among the items you can choose from when defining graphs and tables.

The following steps explain how you can define a multiple-response set, but not how you can use one — that comes later when you generate a table or a graph. Also, there are two Multiple Response menus: The one in the Data menu is for tables and graphs; the one in the Analyze menu is for using special menus that you see in this example.

A multiple-response set can contain a number of variables of various types, but it must be based on two or more *dichotomy variables* (variables with just two values — for example, yes/no or 0/1) or two or more *category variables* (variables with several values — for example, country names or modes of transportation). For example, suppose you have two dichotomy variables with the value 1 defined as "no" and the value 2 defined as "yes." You can create a multiple-response set consisting of all the cases where the answer to both is "yes," where the answer to both is "no," or whatever combination you want.

Do the following to create a simple multiple-response set:

1. **Open the `Apples and Oranges.sav` dataset.**

   Note four dichotomous variables that have 1 for Yes and 0 for No as their possible answers, as shown in Figure 7-6.

2. **Choose Analyze⇨Multiple Response⇨Define Variable Sets.**

   Your variables appear in the Set Definition area. If you previously defined any multiple datasets, they appear in the list on the right.

3. **In the Set Definition list, select each variable you want to include in your new multiple dataset, and then click the arrow to move the selections to the Variables in Set list.**

4. **In the Variable Coding area, select the Dichotomies option and specify a Counted Value of 1.**

5. **Select a Set Name and (optionally) a Set Label.**

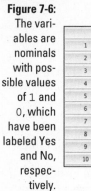

Figure 7-6:
The vari-
ables are
nominals
with pos-
sible values
of 1 and
0, which
have been
labeled Yes
and No,
respec-
tively.

	Apples	Oranges	Pears	Bananas
1	Yes	No	Yes	No
2	Yes	Yes	Yes	No
3	Yes	No	Yes	No
4	Yes	Yes	Yes	No
5	Yes	No	Yes	No
6	Yes	Yes	No	Yes
7	Yes	No	No	Yes
8	Yes	Yes	No	Yes
9	Yes	No	No	Yes
10	No	Yes	No	Yes

6. **Click Add.**

   The new multiple-response set is created and a dollar sign ($) is placed before the name, as shown in Figure 7-7. The dollar sign in the filename identifies the variable as a multiple-response set. The new name will appear in two special menus in the Analyze menu.

   There are other applications of multiple response as well, notably in the menus of the Custom Tables module, but you have to define those multiple-response sets in the Data menu. Having two menus for declaring this variable type can be a little confusing if you don't realize this.

7. **Click Close.**

8. **Choose Analyze⇨Multiple Response⇨Frequencies.**

   The new special variable should appear.

9. **Move the $Fruit variable into the Table(s) For area, as shown in Figure 7-8.**

10. **Move the $Fruit variable into Table(s) For area.**

    The new special Frequencies report appears in the output window, as shown in Figure 7-9.

This may look confusing at first, but it's really pretty easy. Ten people bought 24 pieces of fruit. Nine pieces of fruit were apples — 37.5% of the fruit. Nine out of ten people bought apples — 90% of the people. So, the difference is the denominator: $9/24$ or $9/10$. What makes this table special is that what you usually care about is the people with multiple responses. In other words, how many people shopping at the store are going to buy apples along with other things that they might buy? This table is the only one that easily displays them both ways.

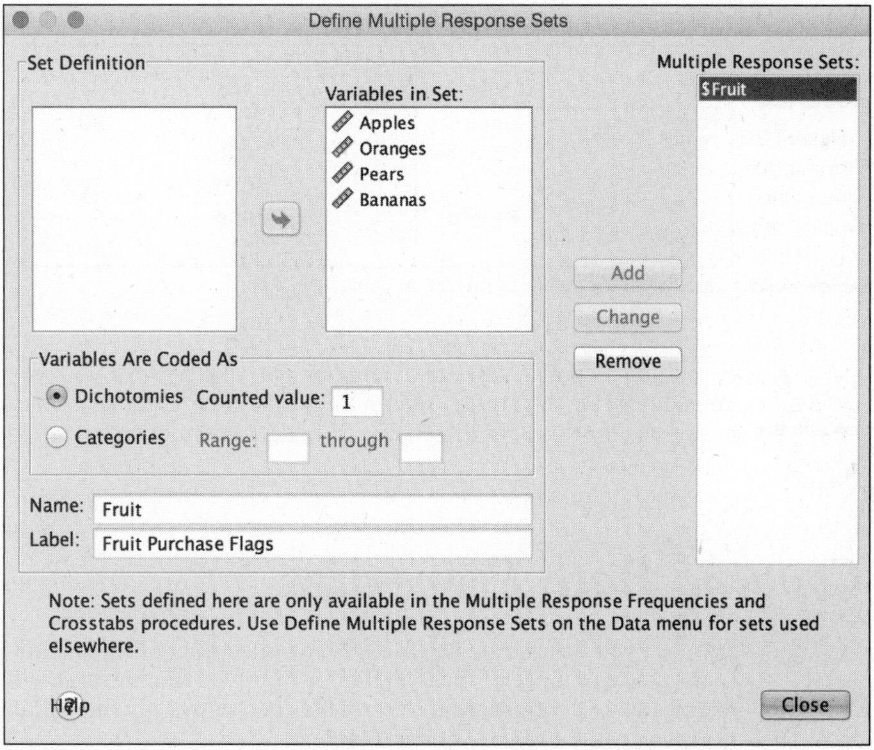

Figure 7-7:
The window showing the complete definition.

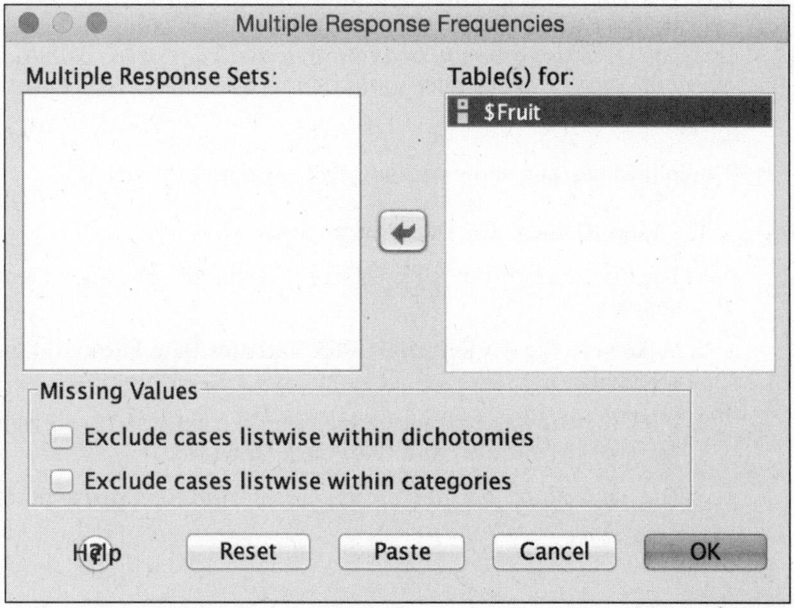

Figure 7-8:
The Multiple Response Frequencies dialog box.

$Fruit Frequencies

		Responses		Percent of Cases
		N	Percent	
Fruit Purchase Flags[a]	Apples	9	37.5%	90.0%
	Oranges	5	20.8%	50.0%
	Pears	5	20.8%	50.0%
	Bananas	5	20.8%	50.0%
Total		24	100.0%	240.0%

Figure 7-9: The Multiple Response Frequencies table.

a. Dichotomy group tabulated at value 1.

If multiple-response sets are a common variable type for you, you should consider trying to get the Custom Tables module because it offers lots of options for this kind of variable. You can read more about modules in Chapter 22.

Copying Data Properties

Suppose you have some data definitions in another SPSS file, and you want to copy one or more of those definitions but you don't want the data. SPSS enables you to choose from several files and copy only the variable definitions you want into your current table.

If you have a variable *of the same name* defined in your table before you execute the copy, you can choose to change the existing variable definition by loading new information from another file. The copied definition simply overwrites the previous information. Otherwise, the copying procedure creates a new variable.

The following steps show you how to copy data properties:

1. **Choose Data➪Copy Data Properties.**

   The Copy Data Properties – Step 1 of 5 window, shown in Figure 7-10, appears.

2. **Make sure the An External SPSS Statistics Data File radio button is selected.**

3. **Click the Browse button, locate the file from which you want to copy variable definitions, and then click Open.**

   The name of the selected file appears next to the Browse button.

4. **Click Next.**

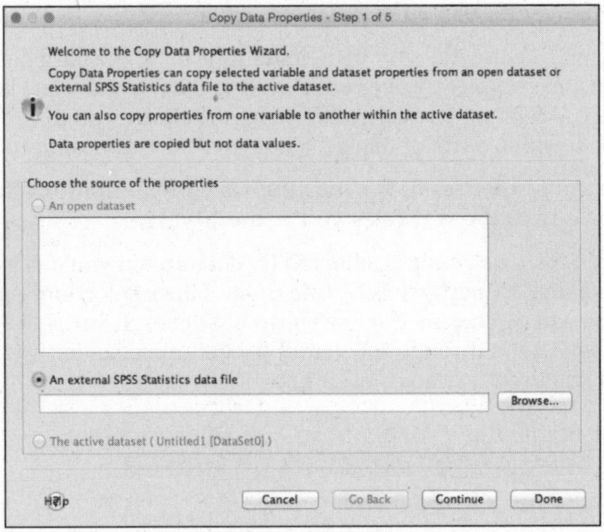

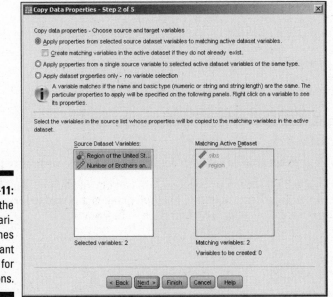

Figure 7-10:
Select the
file you
want to
use as the
source of
variable
definitions.

5. Select the variables you want.

Figure 7-11 displays the variable names that match in the source and destination. In the example, all three are selected, but you can turn the selection of each one on and off. Just put the mouse pointer on the one you want to select or deselect, hold down the Ctrl key, and click.

Figure 7-11:
Select the
source vari-
able names
you want
to use for
definitions.

6. To use the variables you've selected, click Next.

If you want to copy the complete definitions of all the variables you've selected and completely overwrite what you have, you can click Finish. Clicking the Next button, as in this example, allows you to be more specific about which parts of the variable definitions you want to copy.

7. Choose the properties of the existing variable definitions that you want to copy to the variables you're modifying.

In Figure 7-12, everything is selected by default, but you can skip any parts you don't want by deselecting them. These selections apply to all variables you've chosen. If you want to handle each variable separately, you'll have to run through this entire procedure again for each one, selecting different variables each time.

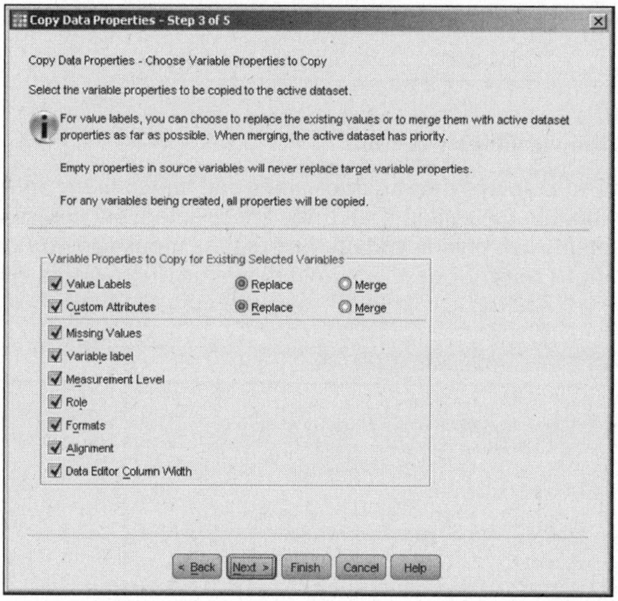

Figure 7-12:
Select which attributes you want to copy.

8. Click Next to be able to select from a list of variable properties.

If you're satisfied with your choices, you can click Finish to complete the process. Clicking Next, as in this example, makes it possible for you to select from a list of available properties to be copied.

9. **Choose any properties made available in the dialog box shown in Figure 7-13.**

   Depending on the variable type, different properties are available to be copied. As shown in Figure 7-13, the properties not available appear grayed out. By default, none of them is selected.

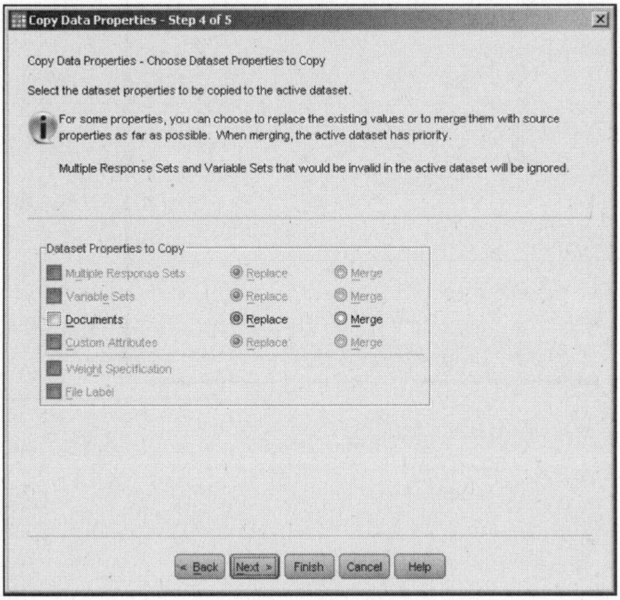

Figure 7-13: Attributes other than variable definitions can be copied from the source.

10. **Click Next to move to the final dialog box.**

    As shown in Figure 7-14, the screen displays the number of existing variable definitions to be changed, the number of new variables to be created, and the number of other properties that will be copied. You can elect to have the action take place immediately or have the set of instructions saved as a Command Syntax script so you can execute them later. (Part VII describes using the Syntax language.)

11. **Decide whether to execute the commands now or later.**

    You can click Finish to have the copy procedure execute immediately.

12. **Click Finish.**

Using the basic variable types and the property descriptions you can add, you should be able to concoct any type of variable you need.

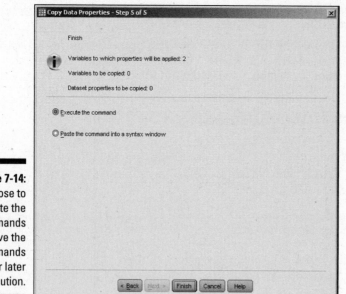

Figure 7-14:
Choose to
execute the
commands
or save the
commands
for later
execution.

Part III
Messing with Data in SPSS

Sort Cases		

Sort by:
- Country of Origin [country_...
- Horsepower [horsepower]...

Sort Order
- ◉ Ascending
- ○ Descending

car_id
Miles per Gallon [mpg]
Engine Displacement (cube...
Vehicle Weight (lbs.) [weight]
Time to Accelerate from 0...
Model Year (modulo 100) [...
American car [american_car]
Number of Cylinders [num...

Save Sorted Data
☐ Save file with sorted data
    File...
☐ Create an index

Help   Reset   Paste   Cancel   OK

Find out about the aggregate procedure in an article at www.dummies.com/extras/spss.

In this part . . .

✔ Transform your data into the form that you need it.

✔ Discover all the best shortcuts and manipulate your data efficiently.

✔ Find out all about the Compute Variable menu, one of the most important menus in SPSS.

✔ Combine data files efficiently.

Chapter 8

The Transform and Data Menus

*A*fter you get your raw data into SPSS, you may find that it contains errors or that it isn't organized the way you'd like. A way to alleviate these problems is by making modifications to your data, configuring the values into a form that's easier to work with and read. This chapter contains some methods you can use to modify your data without losing any information.

A related problem is that you may want to analyze only some of your data, or you may want to perform the analysis more than once. For example, you may want to do a separate analysis for new customers and established customers. You may even want to select the good complete data and avoid the incomplete messy data. It's all about massaging the data after it's in SPSS and making it ready to work for you.

Sorting Cases

You can change the order of your cases (rows) so they appear in just about any order you want. You sort them by comparing the values you entered for your variables. The following example uses the Cars.sav dataset. We sort with two variables, or *sort keys*. The initial sort of the data will simply be by Car ID.

You don't need to limit your sorting to one or two sort keys. You can have a third and fourth key, or more, if necessary, but these keys come into effect only when the keys sorted before them hold identical values. In most cases, two sort keys are plenty to get what you want.

You can sort based on variables of any type simply by selecting the variables as keys. For example:

1. **Choose File ⇨ Open ⇨ Data and open the `Cars.sav` file.**

   The result is the presentation of a collection of apparently unsorted cases shown in Figure 8-1.

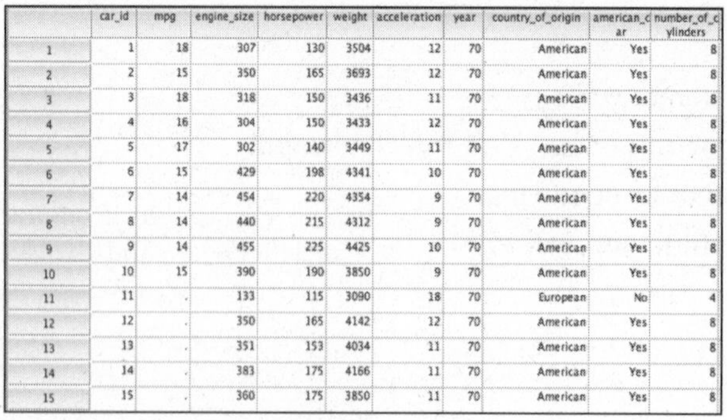

	car_id	mpg	engine_size	horsepower	weight	acceleration	year	country_of_origin	american_car	number_of_cylinders
1	1	18	307	130	3504	12	70	American	Yes	8
2	2	15	350	165	3693	12	70	American	Yes	8
3	3	18	318	150	3436	11	70	American	Yes	8
4	4	16	304	150	3433	12	70	American	Yes	8
5	5	17	302	140	3449	11	70	American	Yes	8
6	6	15	429	198	4341	10	70	American	Yes	8
7	7	14	454	220	4354	9	70	American	Yes	8
8	8	14	440	215	4312	9	70	American	Yes	8
9	9	14	455	225	4425	10	70	American	Yes	8
10	10	15	390	190	3850	9	70	American	Yes	8
11	11	.	133	115	3090	18	70	European	No	4
12	12	.	350	165	4142	12	70	American	Yes	8
13	13	.	351	153	4034	11	70	American	Yes	8
14	14	.	383	175	4166	11	70	American	Yes	8
15	15	.	360	175	3850	11	70	American	Yes	8

Figure 8-1: The data unsorted, as it's loaded directly from the data file.

2. **Choose Data ⇨ Sort Cases.**

   The dialog box is shown in Figure 8-2.

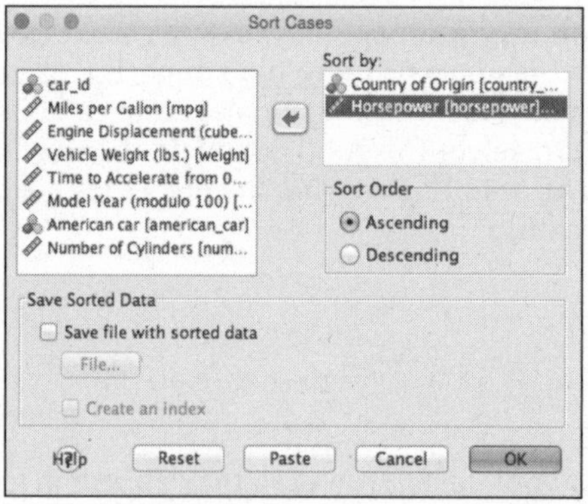

Figure 8-2: The data sorted by horsepower.

3. **Choose the variables Country of Origin and Horsepower, in that order.**

   The result is shown in Figure 8-3.

Figure 8-3:
The data
sorted
alpha-
betically
by Country
of Origin
and then
by Horse-
power.

	car_id	mpg	engine_size	horsepower	weight	acceleration	year	country_of_origin	american_c ar	number_of_c ylinders
248	32	10	360	215	4615	14	70	American	Yes	8
249	102	13	440	215	4735	11	73	American	Yes	8
250	7	14	454	220	4354	9	70	American	Yes	8
251	9	14	455	225	4425	10	70	American	Yes	8
252	20	14	455	225	3086	10	70	American	Yes	8
253	103	12	455	225	4951	11	73	American	Yes	8
254	124	16	400	230	4278	10	73	American	Yes	8
255	338	41	85		1835	17	80	European	No	4
256	362	35	100		2320	16	81	European	No	4
257	26	26	97	46	1835	21	70	European	No	4
258	110	26	97	46	1950	21	73	European	No	4
259	40		97	48	1978	20	71	European	No	4
260	252	43	90	48	1985	22	78	European	No	4

Sorting data is strictly for the way you want it to appear in the table. The order in which the data is displayed never affects the analysis. You can get a quick sense of what's going on by sorting your data, but in the end, it isn't a substitute for a proper analysis in the output window.

The order of the sort keys is important. In the preceding example, if Horsepower had been chosen as the first key and Country of Origin as the second, we would've gotten different results.

If you need to sort using only one variable, you can just right-click the column name.

Selecting the Data You Want to Look At

A very powerful way of manipulating your data is to turn some data "off," while leaving other data "on." In this example, we analyze just European cars, without having to delete anything. SPSS even makes it easy to keep track of what's being counted, averaged, analyzed, and so on, and what's turned "off."

1. **Choose File ⇨ Open ⇨ Data and open the Cars.sav file.**

   If Cars.sav is already open, that's fine, but we'll be starting with the data sorted on Car ID.

2. **Choose Data ⇨ Sort Cases.**

3. **Choose the variable Car ID.**

   Now that the data has been sort on Car ID, we can select the European cars so we can do our analyses just on them.

4. **Choose Data ⇨ Select Cases.**

   The dialog box is shown in Figure 8-4.

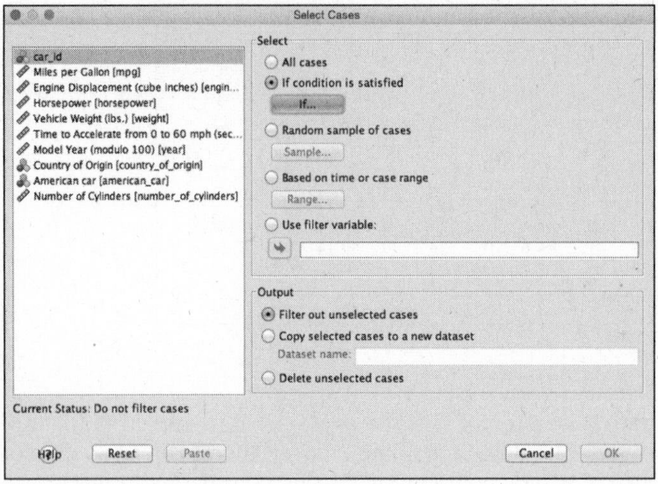

Figure 8-4:
The Select
Cases dia-
log box.

5. Select the If Condition Is Satisfied radio button and then click the If button (refer to Figure 8-4).

You're taken to the dialog box in Figure 8-5.

Now we can specify the selection criteria.

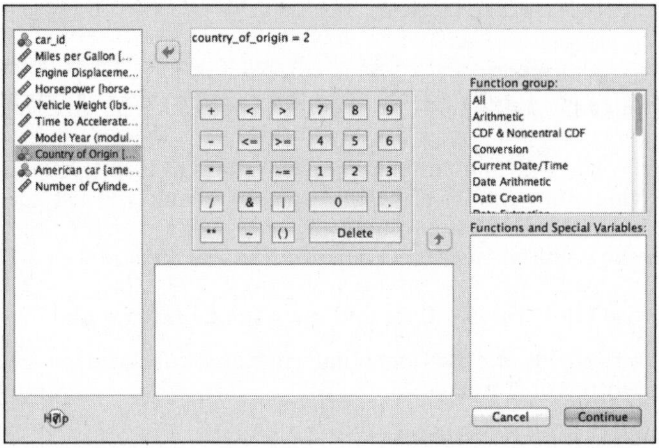

Figure 8-5:
The If dialog
box.

6. Type country_of_origin = 2, **in the expression box and then click OK.**

We have just told SPSS that we want to only select those cases that have a value of 2 on the variable country of origin. It's important that you type the number 2, and not "European" because the actual stored value is 2 and the labeled data is "European."

Figure 8-6 shows the final result. Note the slashes over some of the Row IDs. This shows that the American cars are being ignored (for the time being) and that only the European cars are being analyzed.

Figure 8-6:
The data sorted, indicating selected and unselected cases.

	car_id	mpg	engine_size	horsepower	weight	acceleration	year	country_of_origin	american_car	number_of_cylinders	filter_$
1	1	18	307	130	3504	12	70	American	Yes	8	Not Selected
2	2	15	350	165	3693	12	70	American	Yes	8	Not Selected
3	3	18	318	150	3436	11	70	American	Yes	8	Not Selected
4	4	16	304	150	3433	12	70	American	Yes	8	Not Selected
5	5	17	302	140	3449	11	70	American	Yes	8	Not Selected
6	6	15	429	198	4341	10	70	American	Yes	8	Not Selected
7	7	14	454	220	4354	9	70	American	Yes	8	Not Selected
8	8	14	440	215	4312	9	70	American	Yes	8	Not Selected
9	9	14	455	225	4425	10	70	American	Yes	8	Not Selected
10	10	15	390	190	3850	9	70	American	Yes	8	Not Selected
11	11	.	133	115	3090	18	70	European	No	4	Selected
12	12	.	350	165	4142	12	70	American	Yes	8	Not Selected
13	13	.	351	153	4034	11	70	American	Yes	8	Not Selected
14	14	.	383	175	4166	11	70	American	Yes	8	Not Selected
15	15	.	360	175	3850	11	70	American	Yes	8	Not Selected

From this point forward, every piece of output that you generate will use only the European cars until you turn the Select off. There is a button to return to All Cases in the original menu (refer to Figure 8-4.)

You should always use values and labels for your category values, as is done with the Country of Origin variable in this dataset. This is the way SPSS likes it, and you don't want to make SPSS grumpy, do you? Try typing in just strings, and you're likely to get some errors and random happenings. SPSS's bad mood could soon become your own. Use values and labels to keep everyone happy.

If you wanted to select complete data on a variable (horsepower, for example) you can use the following phrase in the IF formula area: `not(missing(horsepower))`.

Sometimes the values like 1, 2, and 3 are showing in the data window, and sometimes the labels like American, European, and Japanese are showing. There is an easy way to switch back and forth. The button shown in Figure 8-7 appears in the toolbar in the data window. It toggles back and forth between values and labels.

Figure 8-7:
The dataset
toolbar
button for
toggling
between
values and
labels.

Splitting Your Data for Easier Analysis

Under some conditions, you can use an even more powerful version of what we've just illustrated with SELECT. For instance, sometimes you might want to run a series of analyses on one group of cases, and then you can select another group of cases and rerun the same analyses on them. The Split file procedure allows you to select each group in turn, one at a time, and run all your analyses on each separate group.

1. **Choose File ⇨ Open ⇨ Data and open the `Cars.sav` file.**

   If `Cars.sav` is already open, that's fine, but we'll be starting with the data sorted on Car ID. Make sure that the SELECT in the last example has been turned off by returning your SELECT status to All Cases.

2. **Choose Data ⇨ Split File.**

   The dialog box is shown in Figure 8-8.

3. **Choose `Country_of_Origin` as the `Compare Groups` variable and click OK.**

   Your data window won't have slashes as in the case of SELECT. Until we run some output, it won't be clear that anything has changed.

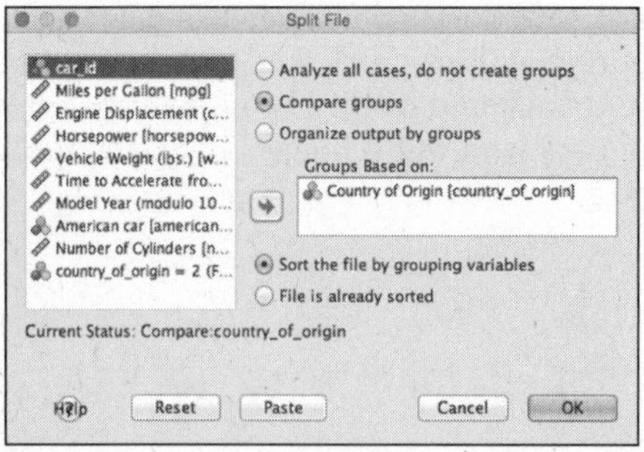

Figure 8-8:
Completed
Split File
dialog box.

4. Choose Analyze ➪ Descriptive Statistics ➪ Frequencies.

5. Choose `Number of Cylinders` **and click OK.**

The resulting output, shown in Figure 8-9, is broken down by Country of Origin. We can stay in this mode as long as we like. Spending hours with a `SPLIT` on is not unheard of when producing tables, charts, and statistics for each of your groups.

Number of Cylinders

Country of Origin			Frequency	Percent	Valid Percent	Cumulative Percent
.	Missing	System	1	100.0		
American	Valid	4	72	28.5	28.5	28.5
		6	74	29.2	29.2	57.7
		8	107	42.3	42.3	100.0
		Total	253	100.0	100.0	
European	Valid	4	66	90.4	90.4	90.4
		5	3	4.1	4.1	94.5
		6	4	5.5	5.5	100.0
		Total	73	100.0	100.0	
Japanese	Valid	3	4	5.1	5.1	5.1
		4	69	87.3	87.3	92.4
		6	6	7.6	7.6	100.0
		Total	79	100.0	100.0	

Figure 8-9: The results of the FREQUENCY while in SPLIT mode.

It's important when you're done with your `SPLIT` (or a `SELECT`) that you turn them off. The option to turn off your `SPLIT` is the Analyze All Cases, Do Not Produce Groups radio button in the original menu shown in Figure 8-8.

In the far bottom right of the data window there is an indicator that tells you whether you currently have a `SPLIT` or `SELECT` operation turned on.

Counting Case Occurrences

If your data is being used to keep track of multiple similar occurrences — such as people who subscribe to any combination of three different magazines, or eggs produced with something other than a single yolk — you can automatically generate a count of the occurrences for each case. SPSS automates the process of creating a new variable and counting the values for you. You specify what value(s) cause a variable to qualify, and SPSS counts the number of qualifying variables from among those you choose. You must have a number of variables that all normally take the same range of values. For example, if you have a number of expenses for each case, you could have SPSS count the number of expenses that exceed a certain threshold.

In the following example, people are listed as subscribers or nonsubscribers to three magazines, which are named simply mag1, mag2, and mag3. The following steps generate a total of the number of subscriptions for each person:

1. **Choose Open ⇨ File ⇨ Data and open the magazines.sav file.**

    This file can be downloaded from the book's companion website at www.dummies.com/go/spss. The screen shown in Figure 8-10 appears.

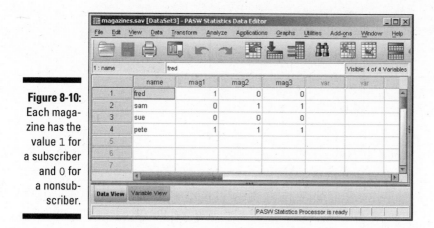

Figure 8-10: Each magazine has the value 1 for a subscriber and 0 for a nonsubscriber.

2. **Choose Transform ⇨ Count Values Within Cases.**

    The screen shown in Figure 8-11 appears.

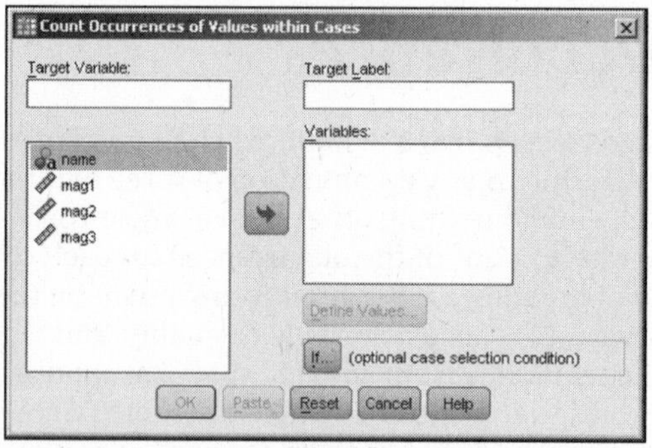

Figure 8-11: The initial value-counting window.

3. **Select the name of every variable you want to use in the count, and then click the arrow to move them from the panel on the left to the panel on the right labeled Variables. Give your new variable a name.**

This operation works only with numerics because it must perform numeric matches on the values. If you want, you can come up with both a name *and* a label to be assigned to the variable that this process creates. In this example, the name is `count` and the label is `Count of subscriptions`, as shown in Figure 8-12.

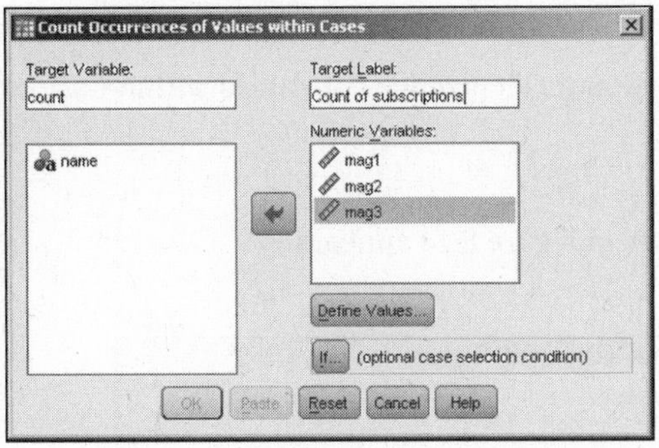

Figure 8-12:
The chosen variables to be counted, and the name of the new variable.

4. Click the Define Values button.

The window shown in Figure 8-13 appears. In this window, we've decided to count, from among the selected variables, those with the numeric value of `1` — which in our example is the value that signifies a subscription.

As you can see in the figure, the total can also be based on missing values and ranges of values. In the ranges, you can specify both the high and low values, or you can specify one end of the range and have the other end be either the largest or the smallest value in the set. In fact, you can select a number of criteria, and SPSS will check each variable against all of them.

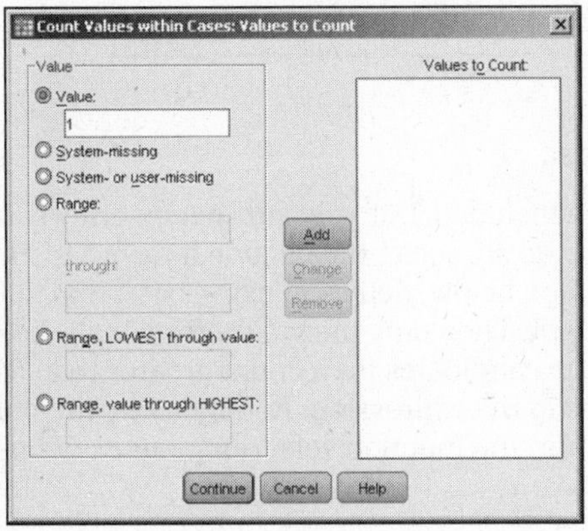

Figure 8-13:
Define the criteria that determine which values are included in the count.

5. **Select a criterion value you want to use, and then click the Add button to move it to the panel on the right labeled Values to Count. Repeat as needed to define all your criteria.**

   The new variable will contain a count of the variables that you named that have a value that matches at least one of the criteria you specified. Each case is counted separately.

6. **Click Continue.**

   You return to the Count Occurrences of Values within Cases screen (refer to Figure 8-11).

7. **Click If.**

   The window shown in Figure 8-14 appears.

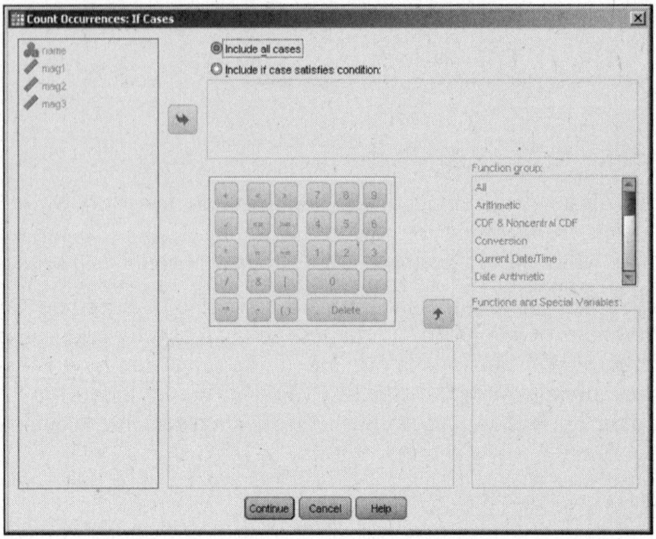

Figure 8-14: Define arithmetic expressions that determine which values are included in the count.

8. **Define your expression.**

   By default, all cases are included, but you can specify criteria here to exclude some cases. To do so, select the Include If Case Satisfies Condition option and, in the text box below, define an expression that specifies the values you want to accept. Then only the values for which the expression is true are considered as candidates for a count greater than 0. You can use any of the variables in the expression. And by using the number pad, the operator buttons, and the function selection, you can construct any expression you want.

9. **Click the Continue button to have SPSS accept your definition. Otherwise (as we did for this simple example), click Cancel and all cases are considered.**

10. **Click the OK button and the new field, along with its counts, is generated.**

The result is the new variable named count, as shown in Figure 8-15.

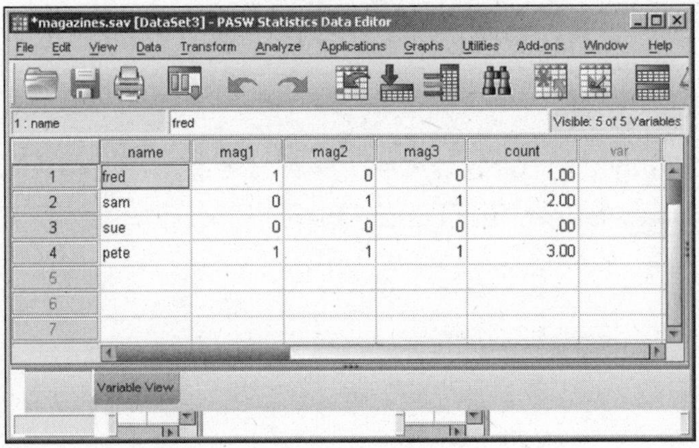

Figure 8-15:
A new variable containing the total number of subscriptions per case.

	name	mag1	mag2	mag3	count	var
1	fred	1	0	0	1.00	
2	sam	0	1	1	2.00	
3	sue	0	0	0	.00	
4	pete	1	1	1	3.00	
5						
6						
7						

Recoding Variables

You can have SPSS change specific values to other specific values according to rules you give it. You can change almost any value to anything else. For example, if you have Yes and No represented by 5 and 6, you could recode the values into 1 and 2. You can recode the values in place without creating a new variable, or you can create a new variable and recode values into it. You may want to do this to correct errors or to make the data easier to use.

When you're recoding values without creating a new variable to receive the new numbers, be sure you store a safety copy of your data before you start. Changes to your data can't be automatically reversed; you could destroy information. For this reason, avoid Recode into Same Variables unless you're sure that you want to use it. The main reason to consider it is if you want to change a bunch of variables all at once. Better to stick with Recode into Different Variables.

Recoding into different variables

Maybe you don't want to overwrite the existing values, but you'd like to have the recoded data available. This is always a safe way to recode. You can always delete the original later if you don't need it. The following steps create the recoded values and are stored in a new variable:

1. **Load the `rsvp.sav` dataset, as shown in Figure 8-16, and choose Transform ⇨ Recode into Different Variables.**

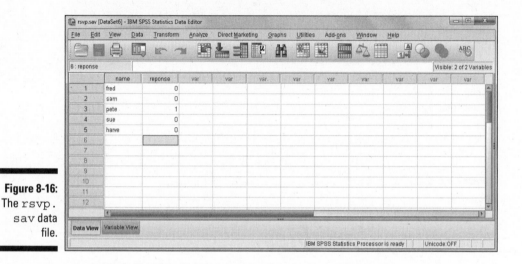

Figure 8-16:
The `rsvp.sav` data file.

2. **In the left panel, select the Response variable holding the values you want to change. Using the arrow in the center, move the variable name to the panel in the center.**

3. **On the right, in the Output Variable area, enter a name** (attending) **and label** (Attending or not) **for a new variable.**

   For the output variable, you can choose a new variable name (so a new variable is created) or choose an existing variable name and have its values overwritten.

4. **Click the Change button and the output variable is defined, as shown in Figure 8-17.**

5. **Click the Old and New Values button.**

6. Define the recoding.

Enter an existing value into the Old Value text box and the value you want it to become in the New Value text box. Then click the Add button to add them to the Old–>New list (as shown in Figure 8-18). Be sure to map all values — even the ones that don't change — because you're creating a new variable and it has no preset values.

7. Click Continue.

8. Click OK.

The results appear, as shown in Figure 8-19. Notice that the numbers all have two digits to the right of the decimal point. This may or may not be what you want, but the new variable was created automatically, and that's part of the default.

Automatic recoding

Automatic recoding converts values into something you can use in computations. For example, if you have a list of automobile names, automatic recoding converts those names into numbers so you can perform an analysis on the pattern of numbers. Automatic recoding gives you a numeric handle on data that could otherwise elude analysis.

To perform automatic recoding, you select options and set the names in a single dialog box. To see an example of automatic recoding in action, follow these steps:

1. **Load `rsvp.sav` (refer to Figure 8-16).**

2. **Choose Transform ⇨ Automatic Recode.**

   The Automatic Recode dialog box appears.

3. **In the panel on the left, select the name of the variable you want to recode. Then click the arrow in the middle to move the variable to the panel on the right.**

4. **In the New Name text box, enter the name of the variable to receive the recoded values.**

5. **Click the Add New Name button.**

   The name you entered appears in the panel above the new name, as shown in Figure 8-20.

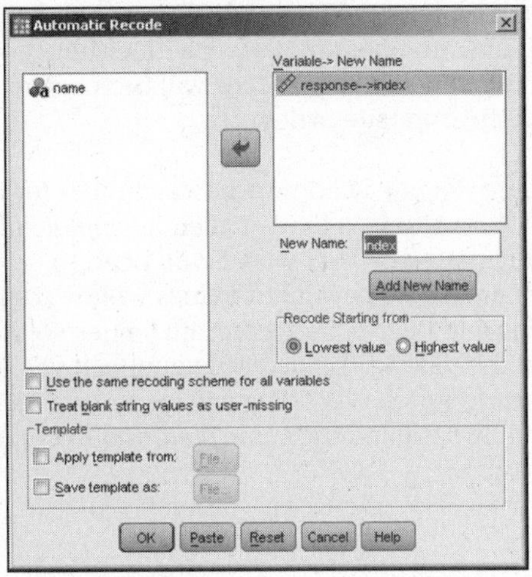

Figure 8-20: The dialog box for automatic recoding.

6. Click OK.

Recoding takes place. The result is similar to that shown in Figure 8-21, where the new variable is named `index`.

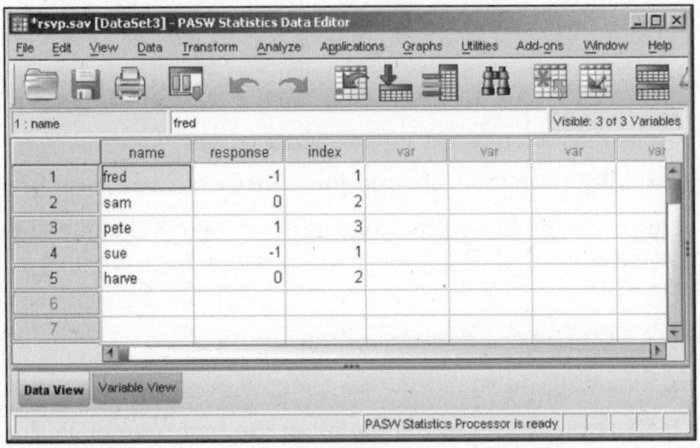

Figure 8-21: The result of automatically recoding `name` into `index`.

The values in the new variable, `index`, come about from sorting the values of the original variable and then assigning numbers to them in that order. If the input values are a string of characters instead of the digits of numbers, the strings are sorted alphabetically (well, almost: uppercase letters come before lowercase).

In the Automatic Recode window (refer to Figure 8-20), you can see the choice for recoding the values with new numbers that start with either the lowest value or the highest value. The new numeric values will be the same either way; they're just assigned in the opposite order.

At the bottom of the Automatic Recode window are two choices for the creation of a template file. This is so you can save a file — called a *template file* — that holds a record of the recoding patterns. That way, if you need to recode more data with the same variable names, the new input values will be compared against the previous encoding and be given appropriate values so that the two data files can be merged and the data will all fit. For example, if you have brand names or part numbers in your data, the recoding will be consistent with the original values because it will be assigned the same *pattern* of recoded values.

Binning

If you're using a scale variable that contains a range of values, you can create groups of those values and organize them into bins. For example, you could use the ages of a number of people and put each one in its own bin — one bin for ages 0 to 20, another bin for ages 21 to 40, and so on. You can specify the size and content of bins in several ways. The actual binning process is automatic.

The following steps take you through an example of the binning process by dividing salaries into bins:

1. **Choose File ⇨ Open ⇨ Data and load the `salaries.sav` file.**

   This file is available for download as described in the introduction. This file contains a list of ID numbers with a salary for each one, as shown in Figure 8-22.

2. **Choose Transform ⇨ Visual Binning.**

   The dialog box shown in Figure 8-23 appears.

3. **Select Current Salary in the panel on the left; then click the arrow in the center of the window to move the name of the variable to the panel on the right.**

4. **Click Continue.**

   A bar graph displaying the range of values of the salaries appears in the center, as shown in Figure 8-24.

5. **Click the Make Cutpoints button.**

   A dialog box appears; here you can specify the size of each bin and the number of bins.

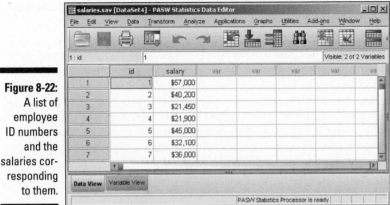

Figure 8-22:
A list of employee ID numbers and the salaries corresponding to them.

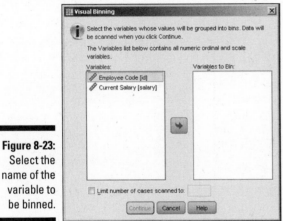

Figure 8-23:
Select the name of the variable to be binned.

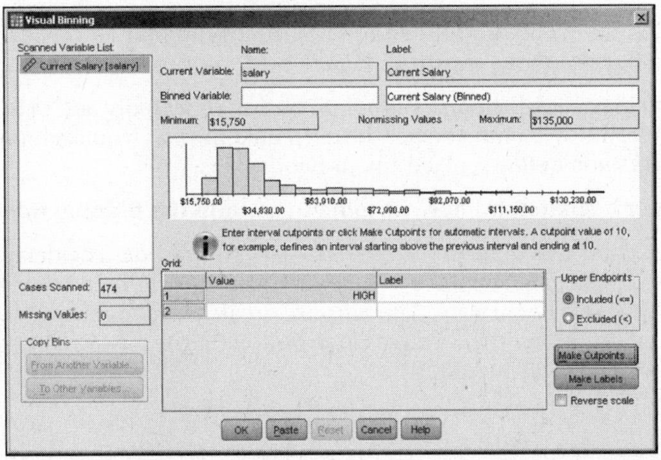

Figure 8-24:
How the binning will be done.

6. Select the points at which you want to have the data cut into parts to create the bins.

In this example, we divided the data into even percentiles of numbers of cases — that is, each bin will contain the same number of cases, as shown in Figure 8-25. Notice that four cutpoints divide the data into five bins, each holding 20% of the cases. We could've chosen to divide the data into equal-width intervals — that is, each bin would contain a range of the same magnitude, which would put different numbers of cases in each bin. Also, the cutpoints could have been based on standard deviations, which would create two cutpoints, dividing the data into the three bins — one each of low, medium, and high capacity.

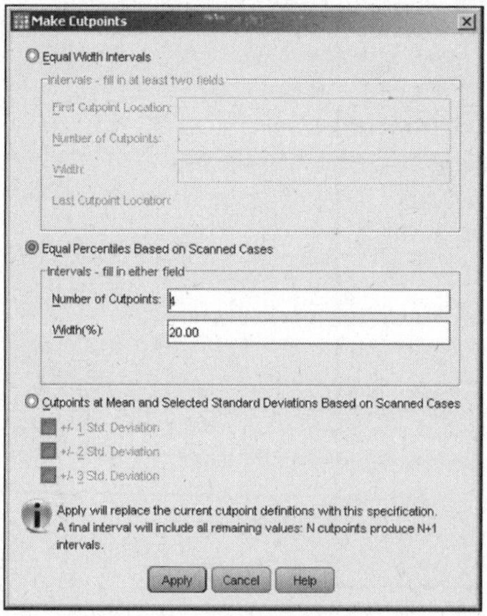

Figure 8-25:
Specify how you want the data divided into bins.

7. Click the Apply button, and the cutpoints appear as vertical lines on the bar graph, as shown in Figure 8-26.

You may click the Make Cutpoints button repeatedly and cut the data different ways until you get the cutpoints the way you like. Any new cutpoints you define replace any previous ones.

8. Enter a name for a new variable to contain the binning information.

You enter the name in the Binned Variable text box. The default label for the new variable appears in the text box to the right of the name. You can change this if you want. The bins are created and numbered from 1 to 5, but if you select the Reverse Scale option (in the lower-right corner), the numbering will be from 5 to 1.

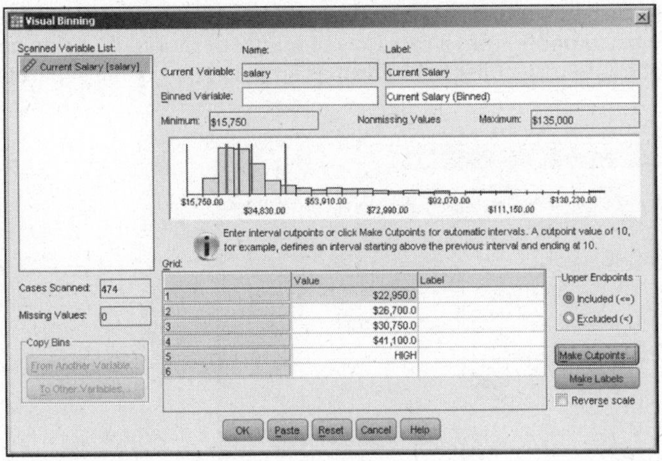

Figure 8-26:
A bar graph
of the data
with cut-
points for
binning.

9. Click OK.

The new variable is created and filled with the bin values, as shown in
Figure 8-27.

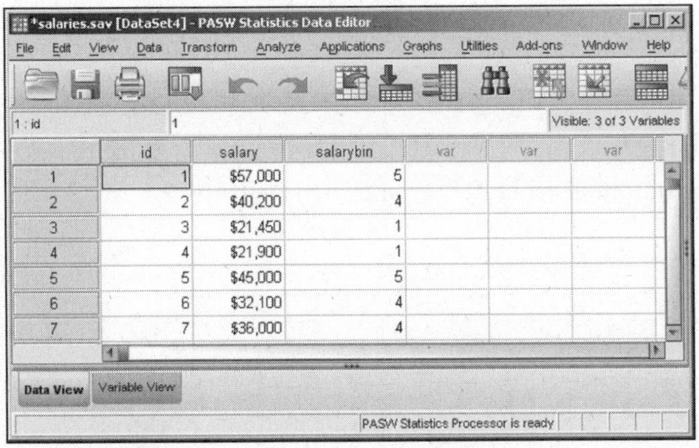

Figure 8-27:
The new
variable
contain-
ing the bin
numbers.

The binning is now complete and you can use the new data for further analysis. One thing you can do quickly and easily is display a summary of the contents of your bins. Simply follow these steps:

1. **With the window in Figure 8-27 still on the screen, choose
   Transform ⇨ Optimal Binning.**

2. **Select variable names on the left and click the arrow buttons to move
   the variables. Move Current Salary to Variables to Bin and move
   Current Salary (binned) to Optimize Bins with Respect To, as shown
   in Figure 8-28.**

The variable in the Optimize Bins with Respect To text box doesn't have to be a variable from a previous binning operation. It can be any variable that contains a collection of values sufficient for being separated into bins.

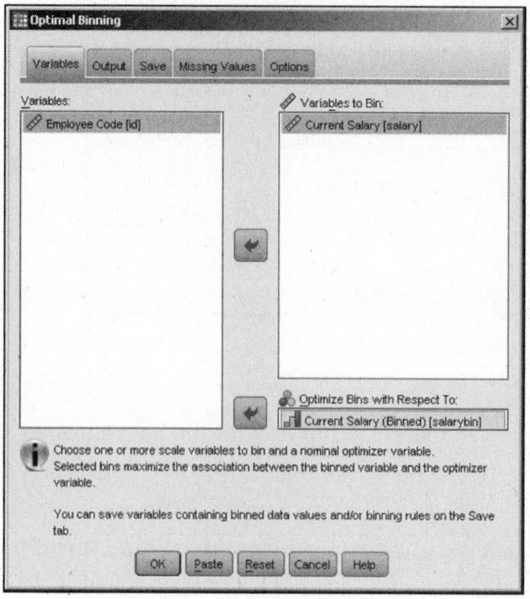

Figure 8-28:
Select the bin variable and the optimizing variable.

3. **Click OK.**

The output is generated, as shown in Figure 8-29.

Current Salary

Bin	End Point		Number of Cases by Level of Current Salary (Binned)					
	Lower	Upper	1	2	3	4	5	Total
1	a	$23,100	96	0	0	0	0	96
2	$23,100	$26,850	0	95	0	0	0	95
3	$26,850	$30,900	0	0	96	0	0	96
4	$30,900	$41,550	0	0	0	93	0	93
5	$41,550	a	0	0	0	0	94	94
Total			96	95	96	93	94	474

Each bin is computed as Lower <= Current Salary < Upper.
a. Unbounded

Figure 8-29:
The output from optimal binning.

Any variable with properly distributed values can be used as the basis of optimal binning. In Figure 8-29, the numbers 1 through 5 across the top are the values of the new binning variable created and stored as part of the data. The numbers 1 through 5 down the left of the graph are the result of the new binning action. The chart lets you see clearly the range of values that make up each bin.

Chapter 9

Using Functions

The Compute Variable dialog box (shown in Figure 9-1) is found in the Transform main menu and is one of the most important dialog boxes in all of SPSS. Why? Inside this dialog box are dozens and dozens of different functions that perform all kinds of different calculations. This is where you go to make new variables out of existing ones. If you don't have much programming experience, it may seem a little tricky, at first. But don't worry — in this chapter, we give you the information you need.

On the right side of Figure 9-1, you see a series of menus with lists and lists of functions. If you use spreadsheet software like Microsoft Excel or Apple Numbers, you may have encountered functions much like these. Here's just a sampling of the kinds of functions that you'll find:

- Arithmetic functions
- Statistical functions
- String functions
- String/numeric conversion functions
- Date and time functions
- Random variable and distribution functions
- Missing value functions
- Logical functions

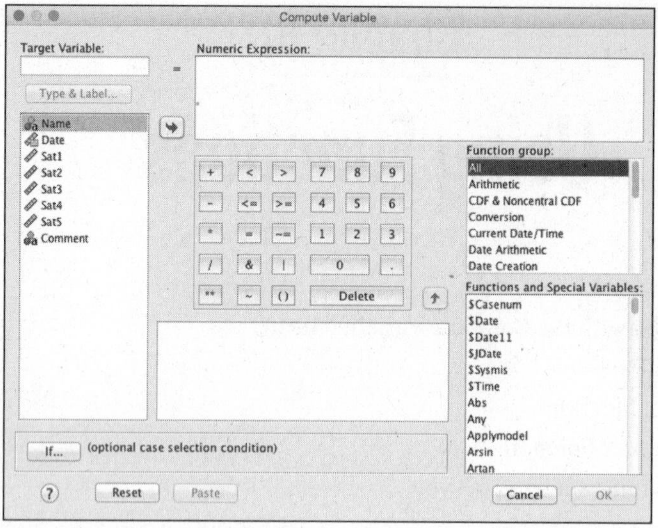

Figure 9-1:
The
Compute
Variable
dialog box.

When you click a function group, like All (refer to Figure 9-1), you see a list of the functions that belong to that function family. In this chapter, we investigate just three of these functions, but a diverse three. The idea is to give you a sense of the different kinds of things you can do, and to introduction you to the function help. For all three examples, we use the `Web Survey.sav` dataset.

The LENGTH Function

At the end of the Web Survey in `Web Survey.sav`, there is an open-ended comment. These kinds of questions often prompt a generic answer like "Not at this time" or "None," but sometimes the responses can be quite useful. The goal is to identify, as simply as possible, if the respondent provided a comment at all. That way, we won't clog up our report with lots of blank rows. When analyzing the answers, we can easily select only those folks who provided a comment. The `LENGTH` function is pretty easy, so it's a good place to start.

1. **Open the `Web Survey.sav` dataset.**

   You may stumble upon `Web Survey 2.sav`. That's what the dataset should look like at the end of the chapter.

2. **Choose Transform ➪ Compute Variable.**

3. **Choose String in the Function Group area in the upper right, and then choose the `LENGTH` function.**

4. **Either drag `LENGTH` into position in the editing area at the top, or click the up-arrow button on the right.**

   The window should look like Figure 9-2.

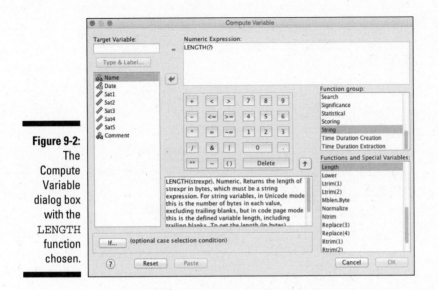

Figure 9-2:
The
Compute
Variable
dialog box
with the
LENGTH
function
chosen.

5. **Either drag Comment into position, replacing the question mark, or click the right-arrow button on the left.**

   Making progress. You could just click OK, but let's take a moment to read the function Help at the bottom of the screen. This is visible whenever you click the Function name (as in Figure 9-2). Onscreen you have to scroll to read it, but the complete text is as follows:

   > LENGTH. LENGTH(strexpr). Numeric. Returns the length of strexpr in bytes, which must be a string expression. For string variables, in Unicode mode this is the number of bytes in each value, excluding trailing blanks, but in code page mode this is the defined variable length, including trailing blanks. To get the length (in bytes) without trailing blanks in code page mode, use LENGTH(RTRIM(strexpr)).

   Whoa! That's a handful, but it all means something. The first word is the name of the function. The second part — LENGTH(strexpr) — is not being repetitive. It's trying to show you the grammar of the command. You have to give it a string expression in parentheses.

   Next, the word *Numeric* means that what it's going to give back to you is, well, a numeric. This little pattern is repeated in every single function. The rest, in the form of sentences, isn't always easy to read, but the first little part is half the battle.

   So, in this case, we give it a word, and it gives us a number. Let's try it.

6. **Give the new variable the new name Length and verify that your screen looks like Figure 9-3.**

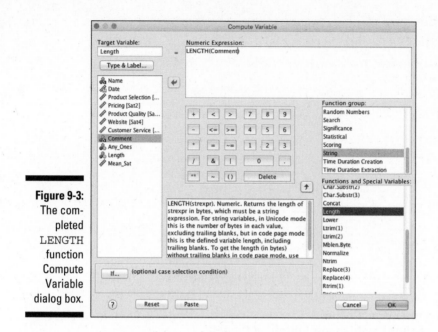

Figure 9-3:
The completed
LENGTH
function
Compute
Variable
dialog box.

7. Click OK.

In the data window, you should find a new variable, Length, listing several numbers. Glance at the original Comment variable. Doug, Jack, and
Georgia did not leave comments, so they should have a zero value.

The ANY Function

When analyzing data, it's often useful to identify people who meet certain
criteria. For example, you may want to identify employees who have high
performance ratings (to keep them in mind for future job openings) or you
may want to identify customers who gave your company a low satisfaction
rating (to see how you can improve their experience). The ANY function
allows you to identify cases that meet the criteria you specify across a series
of variables.

1. Open the Web Survey.sav dataset.

2. Choose Transform ➪ Compute Variable.

**3. Choose Search in the Function Group area in the upper right, and
then choose the ANY function.**

4. **Either drag** ANY **into position in the editing area at the top, or click the up-arrow button on the right.**

The window should look like Figure 9-4.

5. **Review the Function definition at the bottom of the menu.**

Let's take a moment to read the function Help for ANY at the bottom of the screen:

> ANY(test,value[,value,. . .]). Logical. Returns 1 or true if the value of test matches any of the subsequent values; returns 0 or false otherwise. This function requires two or more arguments. For example, ANY(var1, 1, 3, 5) returns 1 if the value of var1 is 1, 3, or 5 and 0 for other values. ANY can also be used to scan a list of variables or expressions for a value. For example, ANY(1, var1, var2, var3) returns 1 if any of the three specified variables has a value of 1 and 0 if all three variables have values other than 1.

Most people don't like these definitions very much, but remember: It all means something. The first word is the name of the function. The second part — ANY(test,value[,value,. . .]) — is warning us that this one is more complicated than LENGTH. It has more moving parts. The next word — *Logical* — tells us that we're going to get a True or False result, which SPSS does with 1 and 0.

For this one, we give it a test value and several variable names, and it gives us a 1 or a 0. Let's try it.

6. **Give the new variable the new name** `Any_Ones`.

7. **Type 1 immediately after the parenthesis in place of the question mark.**

   It looks like you have room for only one more variable because you have only one more question mark, but you can add as many as you like.

8. **Drag all the "Sat" variables (Sat1, Sat2, Sat3, Sat4, and Sat5) with commas separating them just like Figure 9-5.**

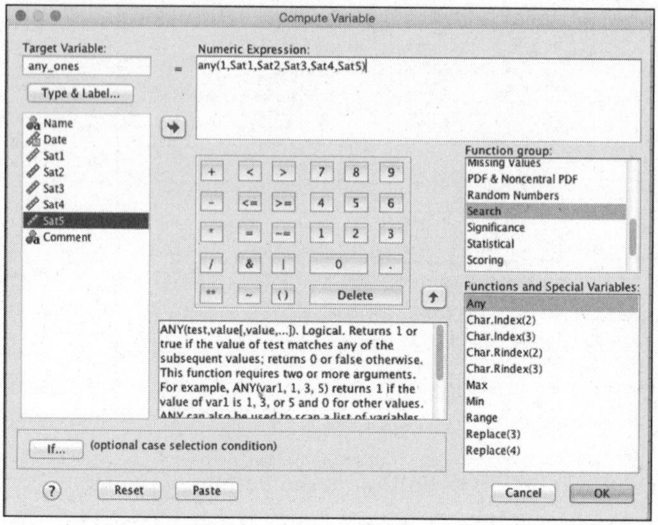

Figure 9-5:
The completed `ANY` function Compute Variable dialog box.

9. **Click OK.**

   In the data window, you should find a new variable, `Any_Ones`, which is `1` (`True`) or `0` (`False`), depending on whether the respondent even gave a `1` (the lowest score). You could easily imagine that management at this company wants to know if anyone is upset enough to give a score of `1` on any question.

The MEAN Function and Missing Data

Now, we're going to calculate the mean of the five satisfaction variables. There is a twist, though — we can include or exclude rows with some missing data. If the respondent didn't answer any of the questions, we don't have much choice, but what if they answer some and not others? That's actually a little tricky. Here's how to do it two different ways:

1. **Open the** `Web Survey.sav` **dataset.**

2. **Choose Transform ⇨ Compute Variable.**

3. **Choose Statistical in the Function Group area in the upper right, and then choose the** MEAN **function.**

4. **Either drag** MEAN **into position in the editing area at the top, or click the up-arrow button on the right.**

5. **Review the function definition at the bottom of the menu.**

   Here's what it says:

   > MEAN. MEAN(numexpr,numexpr[,..]). Numeric. Returns the arithmetic mean of its arguments that have valid, nonmissing values. This function requires two or more arguments, which must be numeric. You can specify a minimum number of valid arguments for this function to be evaluated.

   The first word is the name of the function. The second part — MEAN(numexpr,numexpr[,..]) — is a little easier than LENGTH. We have to provide at least two variables, and three dots tell us we can do more than two, if needed. The next word — *Numeric* — tells us that we're going to get a number as a result.

   We give it several variable names (or numbers), and it gives us a number back.

6. **Give the new variable the new name** Mean_Sat**.**

7. **Drag all the Sat variables (Sat1, Sat2, Sat3, Sat4, and Sat5) with commas separating them just like Figure 9-6.**

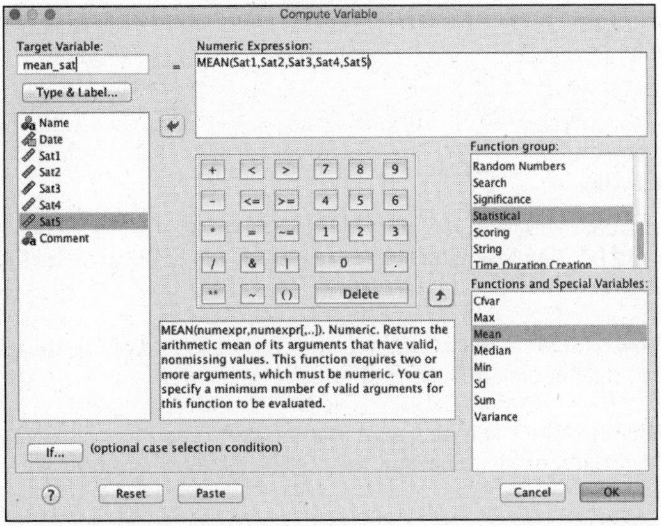

Figure 9-6:
The completed MEAN function Computer Variable dialog box.

8. **Click OK.**

   In the data window, you should find a new variable, `Mean_Sat`, which should be populated for everyone even though Frank didn't answer all the questions. It's basing his average on the answers he did provide.

9. **Return to the Compute Variable menu — our work is still there.**

10. **Give the new variable the new name `Mean_Sat2`.**

11. **Keep all the Sat variables (Sat1, Sat2, Sat3, Sat4, and Sat5) with commas separating them just like Figure 9-6.**

12. **Add two extra characters to the Function so that it reads `MEAN.5`.**

    The `.5` tells SPSS Statistics that in order to compute a mean, each case must provide at least five valid values. It should look like Figure 9-7.

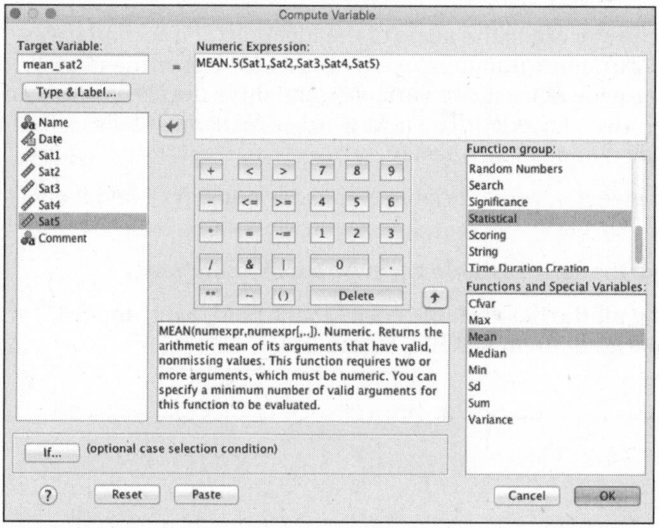

Figure 9-7: The completed MEAN function Compute Variable dialog box.

13. **Click OK.**

    In the data window, you should find a new variable, `Mean_Sat2`, which should be populated for everyone except Frank because he didn't answer all the questions.

This is powerful stuff. Mastering missing values is one of the things that will mark you as an expert in SPSS Statistics.

The finished product can be found in a dataset called `Web Survey 2.sav`. Only one version of Mean Sat appears in the dataset, however.

Chapter 10

Manipulating Files

· ·

· ·

*O*ften data are kept in different files. Sometimes these files are very similar (for example, the same customer information just separated by store); other times the files are very different (for example, customer satisfaction information in one file and financial information in another). IBM SPSS Statistics has two facilities available for merging files: Add Cases joins data files that contain similar variables for separate groups of cases; Add Variables joins data files that contain different information for the same cases.

Merging Files Adding Cases

Add Cases appends cases with the same or similar variables. Figure 10-1 illustrates a simple Add Cases merge of two files containing customer satisfaction records. Both files have the same variables, `Satisfaction` and `Years_Customer`. The variables must have the same name, coded values, and type (for example, string vs. numeric) in both files. In this example, there are three cases in each of the files, so the combined file has six cases. In other words, the combined file contains the total number of cases from both files.

Note that although an ID variable is in both files, it isn't used in an Add Cases merge operation.

ID	Satisfaction	Years_Customer
1	10	5
2	8	7
3	9	7

+

ID	Satisfaction	Years_Customer
101	7	1
102	4	0
103	3	0

=

ID	Satisfaction	Years_Customer
1	10	5
2	8	7
3	9	7
101	7	1
102	4	0
103	3	0

Figure 10-1: Add Cases.

The following example uses two of the data files that are installed with IBM SPSS Statistics: `stroke_invalid` and `stroke_valid`.

1. **From the main menu, choose File⇨Open⇨Data and load the `stroke_invalid` file, which is in the IBM SPSS Statistics directory.**

   The data file with 39 variables and 1,183 cases is shown in Figure 10-2. This file doesn't contain information on whether the patient had a stroke.

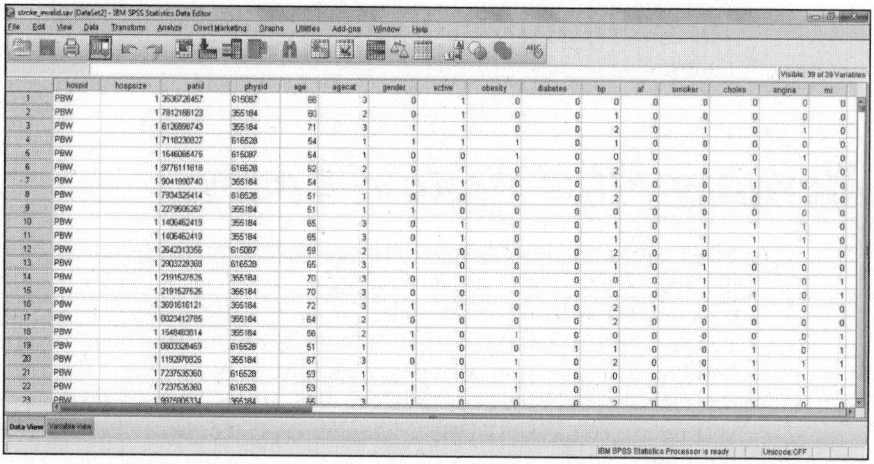

Figure 10-2: The `stroke_invalid` data file.

2. **From the main menu, choose File⇨Open⇨Data and load the `stroke_valid` file, which is in the IBM SPSS Statistics directory.**

   The data file with 42 variables and 1,048 cases is shown in Figure 10-3. This file does contain information on whether the patient had a stroke, which is why it has three additional variables.

   Now we want to combine these two files.

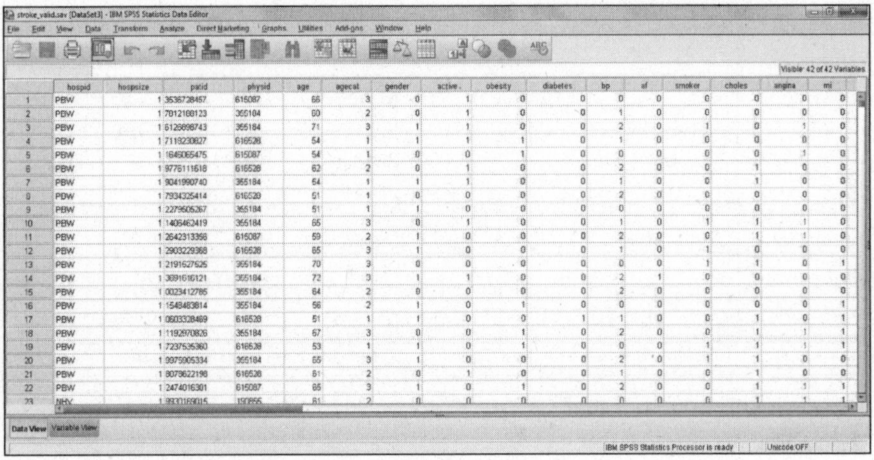

Figure 10-3:
The
`stroke_`
`valid`
data file.

3. **Choose Data ⇨ Merge Files ⇨ Add Cases.**

   The dialog box shown in Figure 10-4 appears.

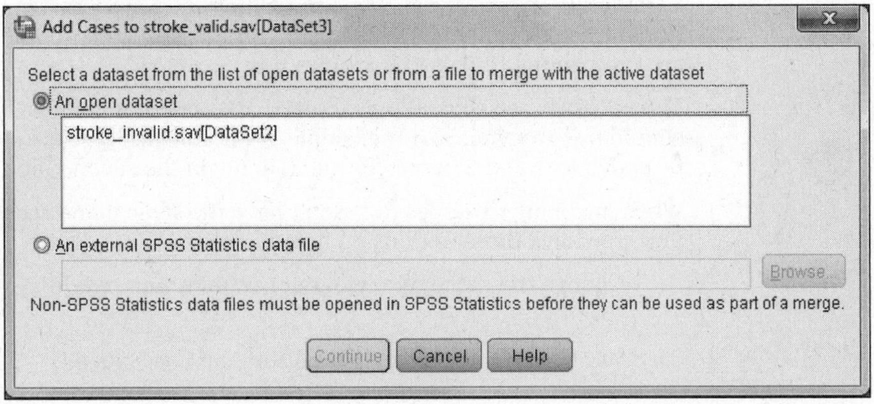

Figure 10-4:
The initial
Add Cases
dialog box.

At this point, you can combine the active dataset (`stroke_valid`) with any files that are open in IBM SPSS Statistics or files that are saved as an IBM SPSS Statistics data file. This means that if you want to combine files in other formats you must first read the files into IBM SPSS Statistics.

4. **Select the file `stroke_valid` and then click Continue.**

   The dialog box shown in Figure 10-5 appears.

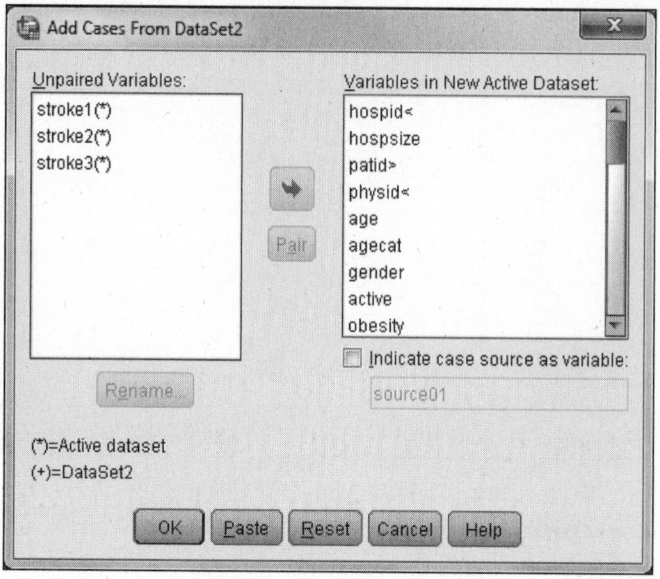

Figure 10-5:
The Add
Cases dia-
log box.

Variables that have the same names in both files are listed in the
Variables in New Active Dataset box. Variables that do not match others
are listed in the Unpaired Variables box.

The variables are aligned by variable name, and the variable formats
should be the same. So, for example, a variable like gender should not
be coded as 1 and 2 in one file and M and F in the second file.

When matching variables that don't have the same name across both
files, you have three options on how to proceed:

- Change the name of one of the variables before adding the files
  together.

- Use the rename option within the Add Cases facility.

- Variables that are not paired can be paired with the Pair button;
  the new variable's name is taken from the variable in the active
  data file.

The file legend in the lower-left corner lists the symbol corresponding to
each file, which is used to designate the source for unpaired variables.

Variables that are unpaired and don't measure the same thing can be
moved to the Variables in New Active Dataset list, and they'll be retained
in the combined file.

5. **Select the variables** `stroke1`, `stroke2`, **and** `stroke3` **and click the button with the arrow to move the variables to the box on the right.**

   The Indicate Case Source as Variable option allows you to create a new variable, named `source01` by default, which will be coded `0` if the case comes from the active dataset and `1` if the case comes from the other data file. This variable can be especially useful if you don't have a variable in the files that uniquely identifies that file.

6. **Select the Indicate Case Source as Variable option and rename the new variable** `file`, **as shown in Figure 10-6.**

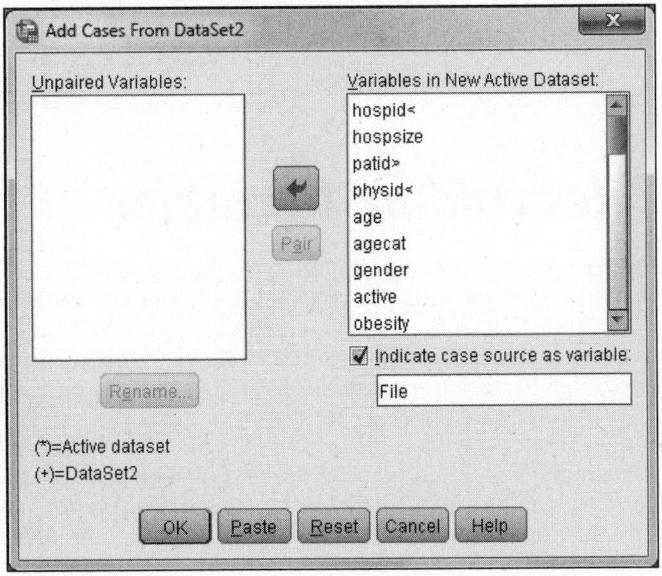

Figure 10-6:
The completed Add Cases dialog box.

Add Cases From DataSet2

Unpaired Variables:

Variables in New Active Dataset:
- hospid<
- hospsize
- patid>
- physid<
- age
- agecat
- gender
- active
- obesity

Pair

☑ Indicate case source as variable:

File

Rename...

(*)=Active dataset
(+)=DataSet2

OK   Paste   Reset   Cancel   Help

7. **Click OK.**

   The new combined file, along with new variable `file` is generated. The result is a new file with 43 variables and 2,231 cases, as shown in Figure 10-7.

Now you can perform analyses on the combined data file, or you can even compare the people in the first file with the people in the second file using the new variable that you just created.

Only two data files can be combined simultaneously when using dialog boxes. However, you can merge an unlimited number of files using syntax.

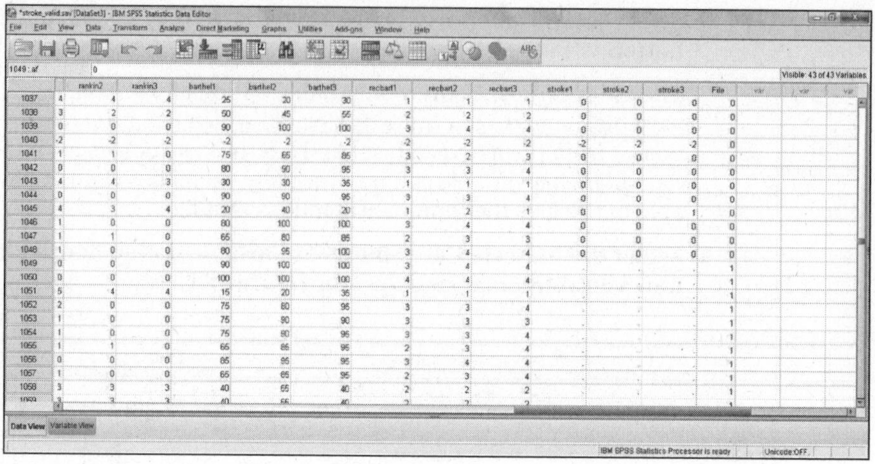

Figure 10-7: The combined data file.

Merging Files Adding Variables

Add Variables joins two data files together so that information held for an individual in different locations can be analyzed together. There are two types of Add Variables merges: one-to-one and one-to-many. Both types add variables to cases matched on key variables. Key variables are case identifiers that exist in both files (for example, a variable like customer ID number).

In one-to-one merges, the basis for the cases is the same in both files and the cases are matched so that one case in the first file corresponds to one case in the second file. In Figure 10-8, for example, `ID` is the key variable used for the match. The resulting file contains all the variables from both files. All cases are retained from both files. Cases not in a file have system-missing values for the variables from that file. In our example, all cases were in both files.

ID	Satisfaction	Years_Customer
1	10	5
2	8	7
3	9	7

+

ID	Product_A	Product_B
1	Yes	No
2	No	Yes
3	Yes	Yes

=

ID	Satisfaction	Years_Customer	Product_A	Product_B
1	10	5	Yes	No
2	8	7	No	Yes
3	9	7	Yes	Yes

Figure 10-8: Adding variables using a one-to-one match.

Both input files must be sorted in ascending order on the key variables to get a one-to-one match to work properly.

In one-to-many merges, one file is designated as the Table file and cases from that file can match to multiple cases in the Case file. The Case file defines the cases in the merged file. The values of the key variable(s) must define unique cases in the Table file, but not in the Case file. In Figure 10-9, each case in the Case file represents a property with information about the property. Each case in the Table file is defined by a zip code with the mean property value for that zip code. Zip_code is the key variable that uniquely identifies each record in this file and is used as the key variable in the one-to-many merge. In the merged file, both cases one and two have the same value for mean_propvalue because they're both in the 85718 zip code.

Case File

PropertyID	Size	No_Rooms	Zip_code
401A	1800	5	85718
502B	2300	5	85718
322C	4200	7	85720
222A	5400	8	85720

Table File

Zip_code	mean_propvalue
85718	232500
85720	785600
85723	425750

Figure 10-9: Adding variables using a one-to-many match.

Zip_code	PropertyID	Size	No_Rooms	mean_propvalue
85718	401A	1800	5	232500
85718	502B	2300	5	232500
85720	322C	4200	7	785600
85720	222A	5400	8	785600

The following example of a one-to-one match uses two of the data files that are not installed with IBM SPSS Statistics: electronics_company_info and electronics_complete.

1. **From the main menu, choose File ⇨ Open ⇨ Data and load the electronics_company_info file, which is not in the IBM SPSS Statistics directory.**

The data file with five variables and 5,003 cases is shown in Figure 10-10. This file contains information on each customer's company.

Input files must be sorted on key variables. In this example the data has already been sorted on the key variable ID.

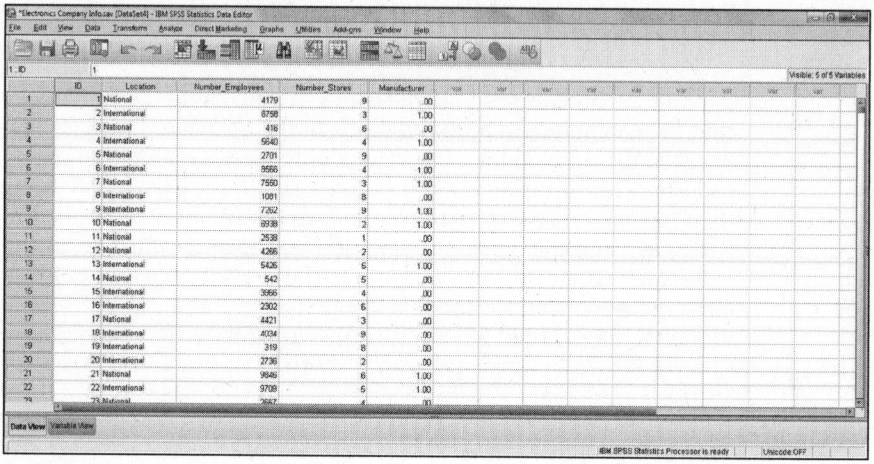

Figure 10-10:
The
electron
ics_com
pany_
info data
file.

2. **From the main menu, choose File ⇨ Open ⇨ Data and load the electronics_complete file, which is not in the IBM SPSS Statistics directory.**

The data file with 12 variables and 5,003 cases is shown in Figure 10-11. This file contains the customer's purchase history.

Input files must be sorted on key variables. In this example, the data has already been sorted on the key variable ID.

Now we want to combine these two files.

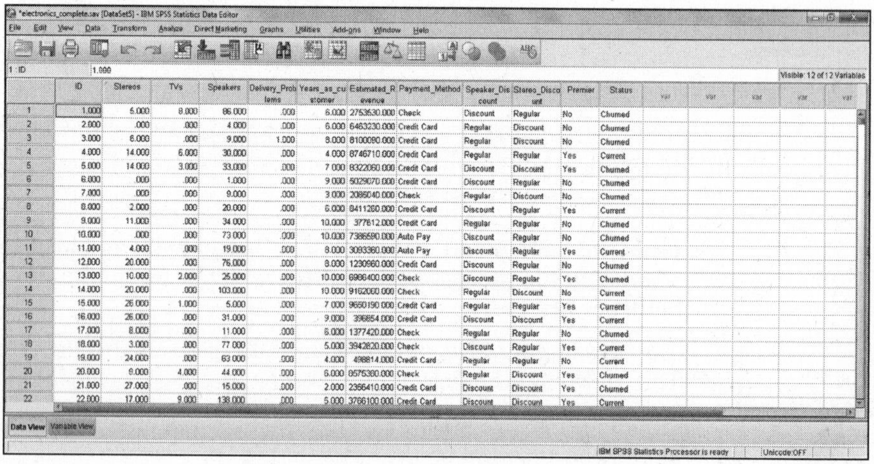

Figure 10-11:
The
electron
ics_com
plete data
file.

3. Choose Data ⇨ Merge Files ⇨ Add Variables.

The dialog box shown in Figure 10-12 appears.

At this point, you can combine the active dataset
(`electronics_complete`) with any files that are open in IBM SPSS
Statistics or files that are saved as an IBM SPSS Statistics data file. This
means that if you want to combine files in other formats, you must first
read the files into IBM SPSS Statistics.

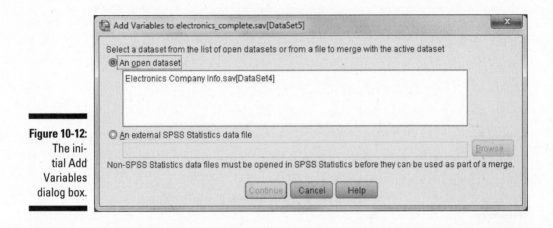

Figure 10-12:
The ini-
tial Add
Variables
dialog box.

4. Select the file `electronics_company_info` and then click Continue.

The dialog box shown in Figure 10-13 appears.

Variables that have unique names are listed in the New Active Dataset
box. If the same variable name is used in both files, only one set of data
values can be retained — these variables will appear in the Excluded
Variables box. Although a renaming facility is available in the Add
Variables dialog box, it's safer to use unique names from the beginning.

If you're merging two files from two time periods, some of the variables
may well have the same name because they measure the same concept.
In this case, each variable should be given a unique name — perhaps
numbered, or based on the date of the survey — to differentiate them.

There are two options to rename variables:

- Change the name of one of the variables before adding the files
together.

- Use the rename option within the Add Cases facility.

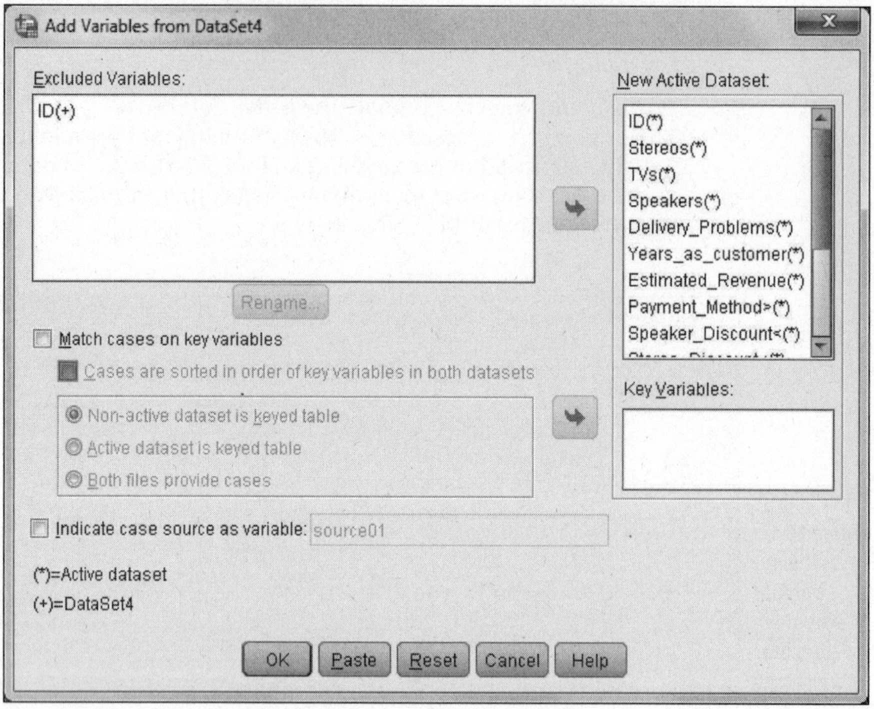

Add Variables from DataSet4

Excluded Variables:

ID(+)

Rename..

☐ Match cases on key variables

☐ Cases are sorted in order of key variables in both datasets

○ Non-active dataset is keyed table
○ Active dataset is keyed table
○ Both files provide cases

☐ Indicate case source as variable: source01

New Active Dataset:

ID(*)
Stereos(*)
TVs(*)
Speakers(*)
Delivery_Problems(*)
Years_as_customer(*)
Estimated_Revenue(*)
Payment_Method>(*)
Speaker_Discount<(*)

Key Variables:

(*)=Active dataset
(+)=DataSet4

OK   Paste   Reset   Cancel   Help

Figure 10-13:
The Add
Variables
dialog box.

Notice that you can click OK at this point. If you do, you'll be doing a merge based on Order. This joins the first record in the first dataset with the first record in the second dataset, and so on. When any of the datasets run out of records, no further output records are produced. This method can be dangerous if there happen to be any cases that are missing from a file or if files have been sorted differently.

5. Select the Match Cases on Key Variables option. Select the variable ID(+), and click the button with the arrow to move the variable to the Key Variables box.

Notice that you can click OK at this point. If you do, you'll be doing a merge based on a left outer join. It's important to note that in a left outer join merge, cases are included only if they have key values that match key values in the active data file.

6. Select the Cases Are Sorted in Order of Key Variables in Both Datasets option. Make sure that the Both Files Provide Cases option is selected (this is the option that produces a one-to-one match), as shown in Figure 10-14.

All the additional options in this dialog box are the same as those previously explained for the merge files Add Cases procedure.

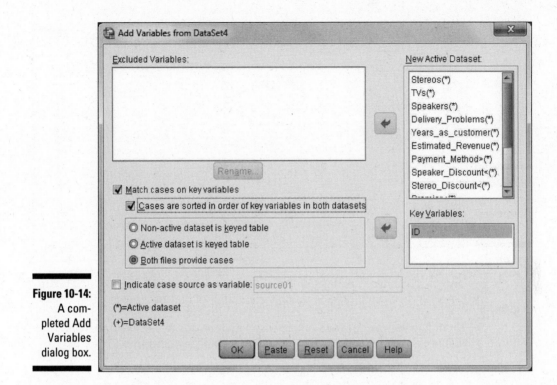

Figure 10-14:
A completed Add Variables dialog box.

7. Click OK.

Figure 10-15 provides a caution that both files must be sorted on the key variable(s). This warning box will always appear and does not indicate that there is a problem with the match.

8. Click OK.

The new combined file is generated. The result is a new file with 16 variables and 5,003 cases, as shown in Figure 10-16.

Now you can perform analyses on the combined data file and investigate relationships that wouldn't have been possible without first performing the merge.

IBM SPSS Statistics 23

Warning: Keyed match will fail if data are not sorted in ascending order of Key Variables.

OK   Cancel

Figure 10-15:
The merge warning.

	ID	Stereos	TVs	Speakers	Delivery_Problems	Years_as_customer	Estimated_Revenue	Payment_Method	Speaker_Discount	Stereo_Discount	Premier	Status	Location	Number_Employees	Num
1	1.000	5.000	6.000	86.000	.000	6.000	2753530.000	Check	Discount	Regular	No	Churned	National	4179	
2	2.000	.000	.000	4.000	.000	6.000	6463230.000	Credit Card	Regular	Discount	No	Churned	International	6758	
3	3.000	6.000	.000	9.000	1.000	8.000	8100090.000	Credit Card	Regular	Discount	No	Churned	National	416	
4	4.000	14.000	6.000	30.000	.000	4.000	8746710.000	Credit Card	Regular	Regular	Yes	Current	International	5640	
5	5.000	14.000	3.000	33.000	.000	7.000	6322060.000	Credit Card	Discount	Discount	Yes	Churned	National	2701	
6	6.000	.000	.000	1.000	.000	9.000	6029070.000	Credit Card	Discount	Regular	No	Churned	International	8566	
7	7.000	.000	.000	9.000	.000	3.000	2085040.000	Check	Regular	Discount	No	Churned	National	7560	
8	8.000	2.000	.000	20.000	.000	6.000	8411260.000	Credit Card	Discount	Regular	Yes	Current	International	1081	
9	9.000	11.000	.000	34.000	.000	10.000	377612.000	Credit Card	Regular	Regular	No	Churned	International	7262	
10	10.000	.000	.000	73.000	.000	10.000	7386590.000	Auto Pay	Discount	Regular	No	Churned	National	6938	
11	11.000	4.000	.000	19.000	.000	8.000	3093360.000	Auto Pay	Discount	Regular	Yes	Current	National	2538	
12	12.000	20.000	.000	76.000	.000	8.000	1230960.000	Credit Card	Discount	Regular	No	Churned	National	4286	
13	13.000	10.000	2.000	25.000	.000	10.000	6986400.000	Check	Discount	Regular	Yes	Churned	International	5436	
14	14.000	20.000	.000	103.000	.000	10.000	9162060.000	Check	Regular	Discount	No	Current	National	542	
15	15.000	26.000	1.000	5.000	.000	7.000	9650190.000	Credit Card	Regular	Regular	Yes	Current	International	3966	
16	16.000	26.000	.000	31.000	.000	9.000	396854.000	Credit Card	Discount	Discount	Yes	Current	International	2302	
17	17.000	6.000	.000	11.000	.000	6.000	1377420.000	Check	Regular	Regular	No	Churned	National	4421	
18	18.000	3.000	.000	77.000	.000	5.000	3942820.000	Check	Discount	Regular	Yes	Current	International	4034	
19	19.000	24.000	.000	63.000	.000	4.000	498814.000	Credit Card	Regular	Regular	No	Current	International	319	
20	20.000	9.000	4.000	44.000	.000	6.000	8575380.000	Check	Regular	Discount	Yes	Churned	International	2736	
21	21.000	27.000	.000	15.000	.000	2.000	2356410.000	Credit Card	Discount	Discount	Yes	Churned	National	9846	
22	22.000	17.000	9.000	138.000	.000	5.000	3766100.000	Credit Card	Discount	Discount	Yes	Current	International	9709	

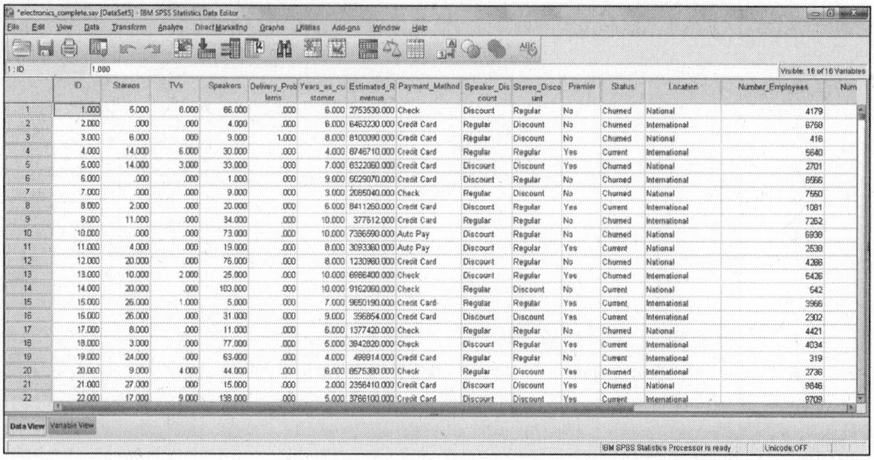

Figure 10-16: The combined data file.

TECHNICAL STUFF

Only two data files can be combined simultaneously when using dialog boxes. However, you can merge an unlimited number of files using syntax.

The following example of a one-to-many match uses two of the data files that are not installed with IBM SPSS Statistics: `rfm_transactions1` and `rfm_aggregated`.

1. **From the main menu, choose File ➪ Open ➪ Data and load the `rfm_aggregated` file, which is not in the IBM SPSS Statistics directory.**

   The data file with four variables and 995 cases is shown in Figure 10-17. This file contains customer's purchase history, where each row represents a customer.

   WARNING!

   Input files must be sorted on key variables. In this example, the data has already been sorted on the key variable `ID`.

2. **From the main menu, choose File ➪ Open ➪ Data and load the `rfm_transactions1` file, which is not in the IBM SPSS Statistics directory.**

   The data file with five variables and 4,906 cases is shown in Figure 10-18. This file contains customer transactional data, where each row represents a transaction.

   WARNING!

   Input files must be sorted on key variables. In this example, the data has already been sorted on the key variable `ID`.

   Now we want to combine these two files.

3. **Choose Data ➪ Merge Files ➪ Add Variables.**
4. **Select the file `rfm_aggregated` and then click Continue.**

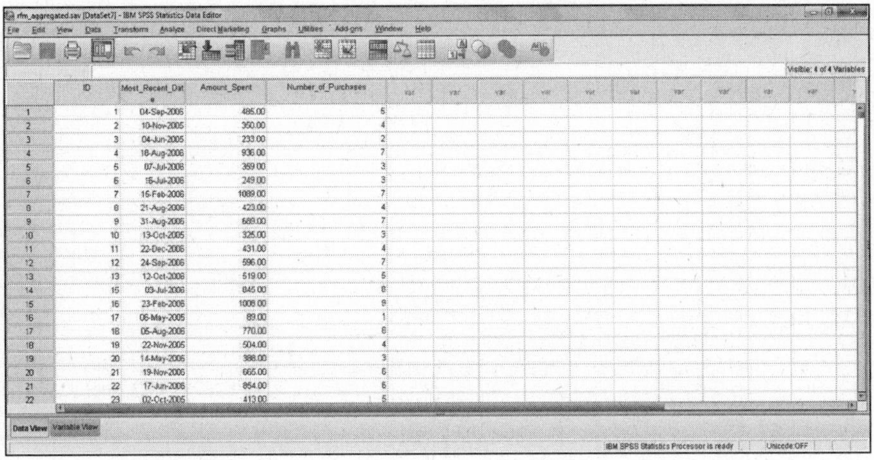

Figure 10-17:
The rfm_
aggre
gated
data file.

5. **Select the Match Cases on Key Variables option. Select the variable id, and click the button with the arrow to move the variable to the Key Variables box.**

6. **Select the Cases Are Sorted in Order of Key Variables in Both Datasets option. Make sure that the Non-active Dataset Is Keyed Table option is selected (this is the option that produces a one-to-many match), as shown in Figure 10-19.**

    The Active Dataset is Keyed Table option is used when doing a many-to-one match.

7. **Click OK.**

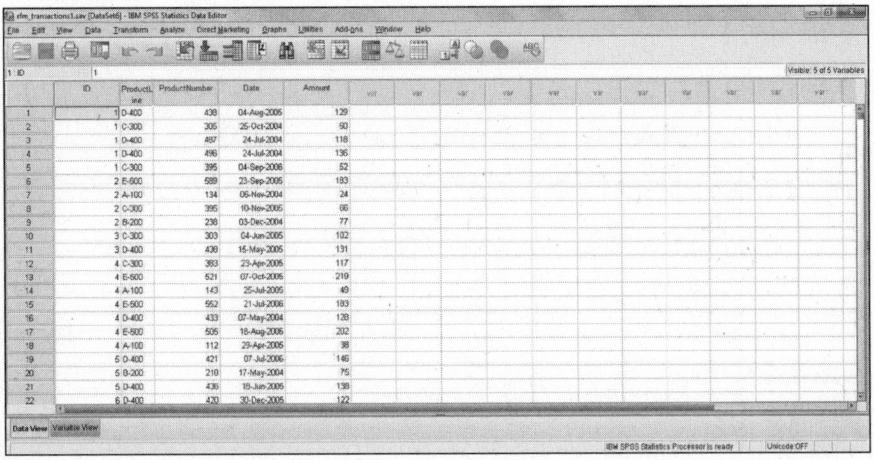

Figure 10-18:
The rfm_
trans
actions1
data file.

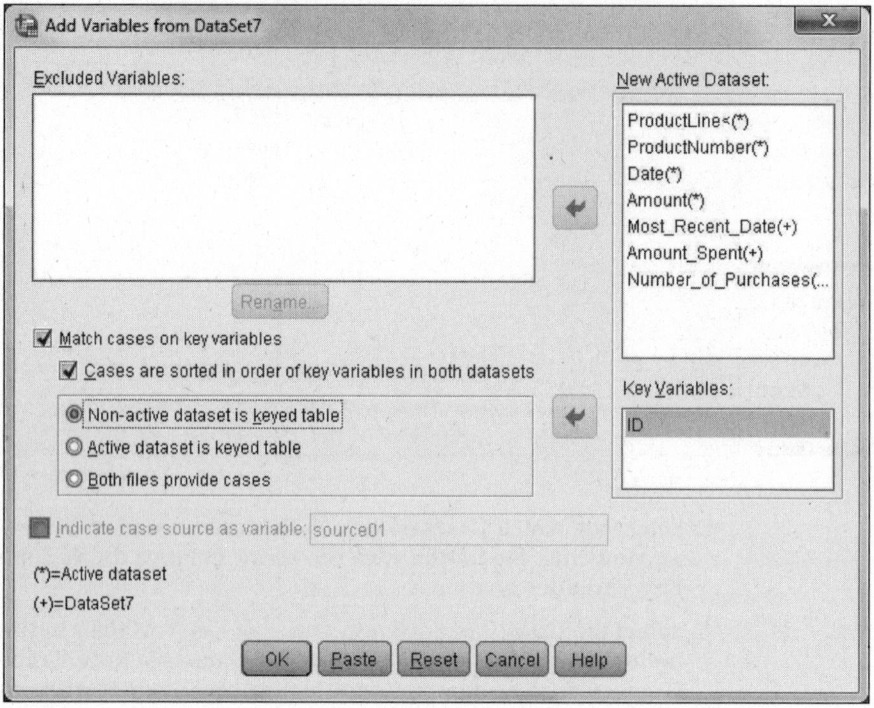

Figure 10-19:
The completed Add Variables dialog box.

8. Click OK again.

> The new combined file is generated, with eight variables and 4,906 cases, as shown in Figure 10-20.

Now you can perform analyses on the combined data file and investigate relationships that wouldn't have been possible without first performing the merge.

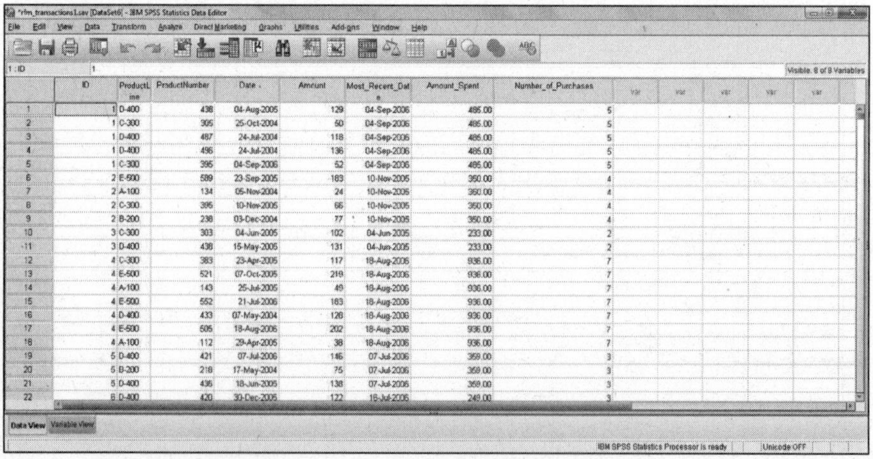

Figure 10-20:
The combined data file.

Part IV
Graphing Data

Chart Builder

Variables:
- AGE OF RESPONDENT [age]
- R'S AGE WHEN 1ST CHILD BORN [agekdbrn]
- WAS R BORN IN THIS COUNTRY [born]
- NUMBER OF CHILDREN [childs]
- SUBJECTIVE CLASS IDENTIFICATION [class]
- YEAR OF BIRTH [cohort]
- R USE COMPUTER [compuse]
- RS HIGHEST DEGREE [degree]
- TYPE OF STRUCTURE [dwelling]
- DOES R OWN OR RENT HOME? [dwelown]
- HIGHEST YEAR OF SCHOOL COMPLETED [educ]
- EMAIL HOURS PER WEEK [emailhr]
- EMAIL MINUTES PER WEEK [emailmin]
- COUNTRY OF FAMILY ORIGIN [ethnic]
- HAPPINESS OF MARRIAGE [hapmar]

No categories (scale variable)

Chart preview uses example data

Drag a Gallery chart here to use it as your starting point

OR

Click on the Basic Elements tab to build a chart element by element

Gallery | Basic Elements | Groups/Point ID | Titles/Footnotes

Choose from:
- Favorites
- Bar
- Line
- Area
- Pie/Polar
- Scatter/Dot
- Histogram
- High–Low
- Boxplot
- Dual Axes

Element Properties...

Options...

Help   Reset   Paste   Cancel   OK

Find out about the Compare Groups graph in a free article at www.dummies.com/extras/spss.

In this part . . .

- ✔ Explore all the graphing options in SPSS.
- ✔ Use Chart Builder to make dozens of different charts.
- ✔ Find the chart types that other people miss.

Chapter 11

On the Menu: Graphing Choices in SPSS

SPSS can display your data in a bar chart, a line graph, an area graph, a pie chart, a scatterplot, a histogram, a collection of high-low indicators, a box plot, or a dual-axis graph. Adding to the flexibility, each of these basic forms can have multiple appearances. For example, a bar chart can have a two- or three-dimensional appearance, represent data in different colors, or contain simple lines or I-beams for bars. The choice of layouts is almost endless. If you really get brave, someday you can learn the programming language behind the graphs called Graphics Production Language (GPL), which truly allows you to do almost anything, but we're getting ahead of ourselves.

In the world of SPSS, the terms *chart* and *graph* mean the same thing and are used interchangeably.

The Graphs menu in the SPSS Data Editor window has three main options: Chart Builder, Graphboard Template Chooser, and Legacy Dialogs (the other options you see are Python plug-ins). These options are different ways of doing the same job. The only reason to do charts with the Legacy Dialogs is if your boss, or professor, or co-workers tell you to. And they're only going to do that if they've been doing it that way for years and changing is too much trouble. In short, don't use the Legacy Dialogs. In this chapter, we work only with the other two.

The Graphboard Template Chooser is a better way of building graphs — and when it hit the scene, the original way of building graphs became known as Legacy. A few years later, an even better procedure for building charts was devised and added to the menu — Chart Builder. All three building methods are in place primarily for people who are in the habit of using the older procedures, but if you build a lot of graphs, you may find advantages and uses for all of them. You can get the same graphs from all three; only the process is different.

Why two "new" menus? The story gets a little complicated. Graphboard Template Chooser can be paired up with an add-on module called SPSS Visualization Designer. (Chapter 22 describes all the modules including this one.) You don't have to own Viz Designer to like Graphboard Template Chooser, however. You get a chance to see it in this chapter. Chart Builder allows you to generate GPL in the Syntax window, which is pretty cool, too. With Version 23, there is another way, but it's very specialized and limited to Spatial Temporal Modeling. It's in its own menu. There's no way around it — they're overlapping systems, and it can get confusing. When in doubt, we recommend Chart Builder, but most graphs can be made both ways.

Building Graphs the Chart Builder Way

SPSS contains Chart Builder, which uses a graphic display to guide you through the steps of constructing your display. It checks what you're doing as you proceed and won't allow you to use things that won't work.

The Gallery tab

The following example steps you through the process of creating a bar chart, but you can use the same fundamental procedure to build a chart of any design. You can follow this tutorial once to see how it all works. Later on, you can use your own data and choices.

You can't hurt your data by generating a graphic display. Even if you thoroughly mess up the graph, you can always redo it without fear. This is one place where mistakes don't cost anything. And nobody's watching.

The following steps build a bar chart:

1. **Choose File⇨Open⇨Data and load the** GSS2012 Abbreviated. sav **file, which is one of the supporting files on the book's website** (www.dummies.com/go/spss).

2. Choose Graphs⇨Chart Builder.

A warning appears, informing you that before you use this dialog box, measurement level should be set properly for each variable in your chart. (We have the correct measurement level so you can proceed.)

3. Click OK.

The Chart Builder dialog box appears, as shown in Figure 11-1. If a graph was generated previously, the display will be different, and you'll need to click the Reset button to clear the Chart Builder display.

4. Make sure the Gallery tab is selected.

5. In the Choose From list, select Bar as the graph type.

The fundamental types of bar charts appear in the gallery to the right of the list.

6. Define the general shape of the bar graph to be drawn.

You can do so in two ways. The simplest is to choose one member from the set of diagrams of bar graphs appearing immediately to the right of the list. For this exercise, select the diagram in the upper-left corner and drag it to the large chart preview panel at the top.

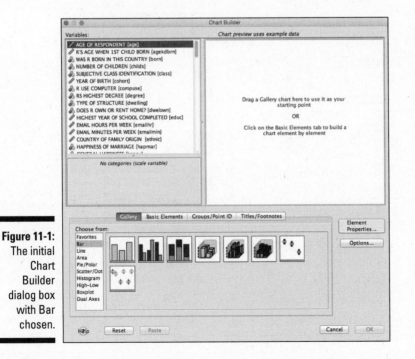

Figure 11-1: The initial Chart Builder dialog box with Bar chosen.

Alternatively, you can click the Basic Elements tab (instead of the Gallery tab) and drag one image from each of the two displayed panels to the panel on top, which constructs the same diagram as the bar graph.

Figure 11-2 shows the appearance of the window after the dragging is complete. The result is the same no matter which procedure you follow.

You can always back up and start over: At any time during the design of a graph, click the Reset button. Anything you dragged to the display panel is deleted, and you can start from scratch.

7. **Click Close to close the Element Properties window (see Figure 11-3).**

   This window should've popped up when you dragged the graphic layout to the panel. This dialog box is not needed for this example, so you can close it. If it didn't appear but you'd like to see it, you can click the Element Properties button at any time.

8. **From the list on the left, select the variable with the label and name Highest Year of School Completed (Educ) and drag it to the Y-Axis label in the diagram.**

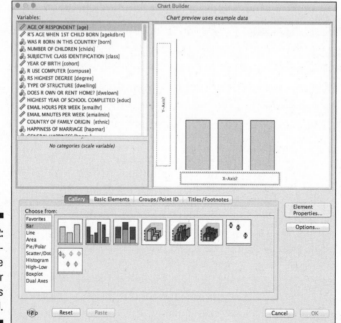

Figure 11-2: The appearance of the new bar chart is defined.

Figure 11-3:
Use the
Element
Properties
window to
modify chart
elements.

9. **In similar fashion, select the variable with the label and name Region of Interview (region) and drag it to the X-Axis label in the diagram.**

   The screen now looks like the one shown in Figure 11-4.

   The graphics display inside the Chart Builder window *never* represents your actual data, even after you insert variable names. This window simply displays a diagram that demonstrates the composition and appearance of the graph that will be produced.

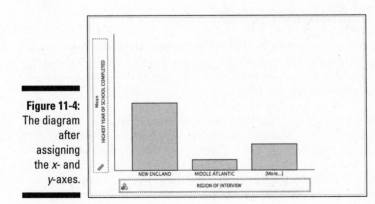

Figure 11-4:
The diagram
after
assigning
the *x*- and
y-axes.

10. Click the OK button to produce the graph.

An SPSS Statistics Viewer window appears, containing the graph shown in Figure 11-5. This graph is based on the actual data; it shows that the average number of years of education varied little from one part of the country to the next in this survey.

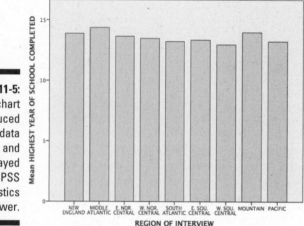

Figure 11-5: A bar chart produced from a data file and displayed by SPSS Statistics Viewer.

These steps demonstrate the simplest way possible of generating a chart. Most of the options available to you were left out of the example so it would demonstrate the simplicity of the basic process. The result could also use a little editing to the *y*-axis, but editing is covered in Chapter 18. The following sections describe the options.

The Basic Elements tab

The example in the preceding section uses the Gallery tab to select the type and appearance of the chart. Alternatively, you can click the Basic Elements tab in the Chart Builder dialog box and select one part of the chart from each of the two panels shown in Figure 11-6.

The Basic Elements tab allows you to choose one element from column A and another from column B. You drag one image from each panel into the panel at the top, and they combine to construct a diagram of the graph you want.

The result is the same as you get from using the Gallery tab. The only difference is that you use the Basic Elements tab to build the graph from its components. Whether you use this technique or the Gallery depends on your conception of the graph you want to produce.

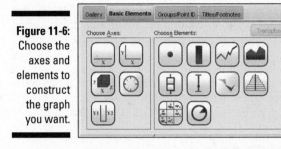

Figure 11-6:
Choose the axes and elements to construct the graph you want.

The Groups/Point ID tab

After you've selected the type and appearance of your chart through either the Gallery tab or the Basic Elements tab, you can click the Groups/Point ID tab in the Chart Builder dialog box, which provides you with a group of options you can use to add another dimension to your graph.

In the example in Figure 11-7, we selected the Rows Panel Variable option, which generates a multifaceted graph. The new dimension adds a separate graph for whether the respondent uses a computer. A separate set of bars is drawn: one for those who use a computer, and one for those that do not.

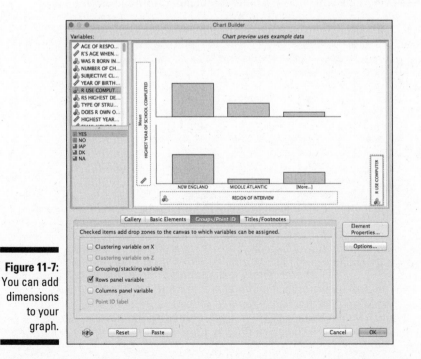

Figure 11-7:
You can add dimensions to your graph.

The Columns Panel Variable option (located in the Groups/Point ID tab of the Chart Builder) enables you to add a variable along the other axis, thus adding another dimension. Adding variables and new dimensions this way is known as *paneling,* or *faceting.*

Clustering (gathering data into groups) can also be done along the *x-* or *y-*axis if the variables are the type that will cluster (or bin) properly.

The Titles/Footnotes tab

Figure 11-8 shows the window you get when you click the Titles/Footnotes tab in the Chart Builder dialog box. Each option in the bottom panel places text at a different location on the graph. When you select an option, the Element Properties window appears so you can enter the text for the specified location.

The Element Properties dialog box

You can use the Element Properties dialog box at any time during the design of a chart to set the properties of the individual elements in the chart. One mode of the dialog box is shown in Figure 11-3; another is shown in Figure 11-9. It changes every time you choose a different member from the list at its top.

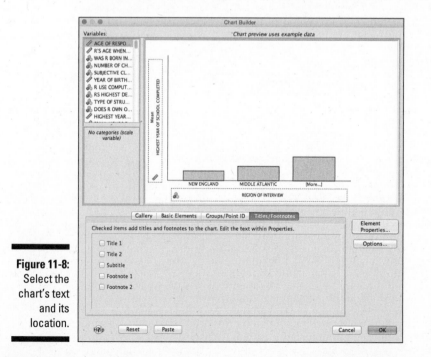

Figure 11-8:
Select the chart's text and its location.

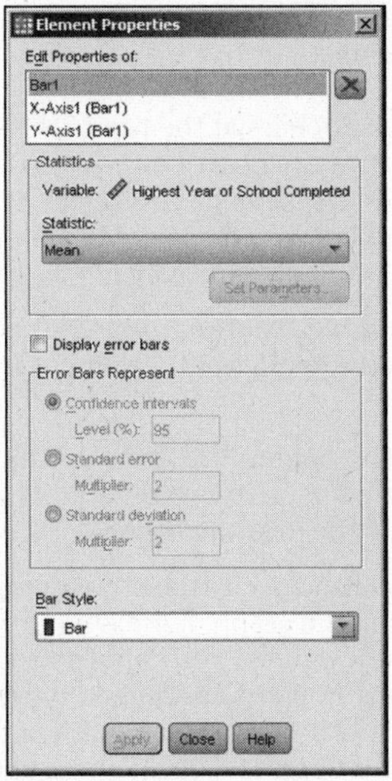

Element Properties

Edit Properties of:

Bar1
X-Axis1 (Bar1)
Y-Axis1 (Bar1)

Statistics

Variable: Highest Year of School Completed

Statistic:

Mean

Set Parameters

☐ Display error bars

Error Bars Represent

⦿ Confidence intervals
  Level (%): 95
◯ Standard error
  Multiplier: 2
◯ Standard deviation
  Multiplier: 2

Bar Style:

Bar

Apply   Close   Help

Figure 11-9:
The options
for an axis
variable.

The dialog box often pops up on its own when you add an item to the graph's
definition. You can make it appear any time you want by clicking the Element
Properties button in the Chart Builder dialog box.

Okay, the upcoming list of options is long, but four facts make them simple
to use:

- ✔ **All options have reasonable defaults.** You don't have to change any of
  them unless you want to.

- ✔ **You can always back up and change whatever settings you made.**
  Nothing is permanent, so you can make changes until you've finished or
  run out of time and decide, "That's good enough."

- ✔ **Not all options appear at once.** Only a few show up at a time. In fact,
  you'll probably never see some of the possible options.

- ✔ **All options become obvious when you see what they do.** You don't
  have to memorize any of them, but you'll find they're easy to remember.

The following is a simple explanation of all the possible options that can appear in an Element Properties dialog box:

- **Edit Properties Of:** This list, which appears at the top of the window, is used for selecting which element in the chart you want to edit. Each element has a type, and the type of the element you select determines the other options available in the window. The selected element is also highlighted in the diagram of the graph in some way.

- **X:** When an element is selected and the X button to the right of the list becomes enabled, clicking the button removes the element from the list and from the graph.

- **Arrow:** For charts with dual y-axis variables, the arrow to the right in the list indicates which of the variables will be drawn on top of the other. You can click the arrows to change the drawing order.

- **Statistic:** For certain elements, you can specify the statistics (that is, the type of value) to be displayed in the graph. For example, you can select Count and use simple numeric values. You can also select Sum, Median, Variance, Percentile, or any of up to 32 statistic types. Not all types of charts have that many options; the options that are available also depend on the types of variables you're using. For certain statistics options — such as Number in Range and Percentage Less Than — the Set Parameters button is activated; you have to click it to set the parameters controlling your choice.

- **Axis Label:** You can change the text used to describe a variable. By default, the variable's label is used.

- **Automatic:** If selected, the range of the selected axis is determined automatically to include all the values of the scale variable being displayed along that axis. This is the default.

- **Minimum/Maximum:** You can replace the Automatic default values and choose the extreme values that determine the starting and ending points of an axis.

- **Origin:** Specifies a point from which chart information is graphed. This option has different effects for different types of charts. (For example, choosing an origin value for a bar chart can cause bars to extend both up and down from a center line.)

- **Major Increment:** The spacing that determines the placing of tick marks, along with numeric or textual labels, on an axis. The value of this option determines the interval of spacing when you also specify minimum and maximum values.

✔ **Scale Type:** You have four different types of scale you can use along an axis:

- **Linear:** A simple, rulerlike scale. This is the default.
- **Logarithmic (standard):** Transforms the values into logarithmic values for display. You can also select a base for the logarithms.
- **Logarithmic (safe):** Same as standard logarithms, except the formulas that calculate values can handle 0 and negative numbers.
- **Power:** Raises the values to an exponential power. You can select an exponent other than the default value of 0.5 (which is the square root).

✔ **Sort By:** You can select which characteristic of a variable will be used as the sort key. It can be one of the following three:

- **Label:** Nominal variables are sorted by the names assigned to the values; you can choose whether to sort in ascending or descending order.
- **Value:** Uses the numeric values for sorting. You can choose whether to sort in ascending or descending order.
- **Custom:** Uses the order specified in the Order list.

✔ **Order:** The list of possible values is flanked by up and down arrows. You can change the sorting order by selecting a value and clicking an arrow to move the selection up or down. To remove a value from the produced chart, select its name in the list and click the X button; the value moves to the Excluded list. When you change the Order list, Sort By switches automatically to Custom.

✔ **Excluded:** Any value you want to exclude from the Order list appears in this list. To move a value back to the Order list (which also causes the value to reappear on the chart), select its name and click the arrow to the right of the list.

If a value (or a margin annotation representing a value) is unexpectedly missing from a graph based your selections, look in this Excluded list. You may have excluded too much.

✔ **Collapse:** If you have a number of values that seldom occur, you can select this option to have them gathered into an "Other" category. You specify the percentage of the total number of occurrences to make it an "Other" value.

✔ **Error Bars:** For Mean, Median, Count, and Percentage, confidence intervals are displayed. For Mean, you must choose whether the error bars will represent the confidence interval, a multiple of the standard error, or a multiple of the standard deviation.

- ✔ **Bar style:** You can choose one of three possible appearances of the bars on a bar graph.

- ✔ **Categories:** You can choose the order in which the values appear when they're placed along an axis. You can select ascending or descending order. If the variable is nominal, you can select the individual order and even specify values to be left out.

- ✔ **Small/Empty Categories:** You can choose to include or exclude missing value information.

- ✔ **Display Normal Curve:** For a histogram, you can choose to have a normal curve superimposed over the chart. The curve will use the same mean and standard deviation values as the histogram.

- ✔ **Stack Identical Values:** For a chart that will appear as a *dot plot* (a pattern of plotted points), you can choose whether points at the same location should appear next to one another or one on top of one another (that is, with one point blotting out the one below it).

- ✔ **Display Vertical Drop Lines between Points:** For a chart that will appear as a *dot plot* (a pattern of plotted points), any points with the same *x*-axis values show a vertical line joining them.

- ✔ **Plot Shape:** For a dot plot, you can choose

  - **Asymmetric:** Stacks the points on the *x*-axis. This is the default.

  - **Symmetric:** Stacks the points centered around a line drawn horizontally across the center of the screen.

  - **Flat:** The same as Symmetric, except no line is drawn.

- ✔ **Interpolation:** For line and area charts, the algorithm used to calculate how the line should be drawn between points:

  - **Straight:** Draws a line directly from one point to the next.

  - **Step:** Draws a horizontal line through each point; the ends of the horizontal lines are connected with vertical lines.

  - **Jump:** Draws a horizontal line through each point, but the ends of the lines are not connected.

  - **Location:** For Step and Jump interpolation; using this option adds an indicator at the actual point.

  - **Interpolation through Missing Values:** For Straight, Step, or Jump, this option draws lines through missing values. Otherwise, the line shows a gap.

- ✔ **Anchor Bin:** The starting value of the first bin. This option is available for histograms.

- ✔ **Bin Sizes:** Sets the sizes of the bins when you're producing a histogram.

✔ **Angle:** Rotates a pie chart by selecting the clock position at which the first value starts. You can also specify whether the values should be included clockwise or counterclockwise.

✔ **Display Axis:** For a pie chart, you can choose to display the axis points on the outer rim.

The Options dialog box

Clicking the Options button in the Chart Builder dialog box opens the Options dialog box, shown in Figure 11-10.

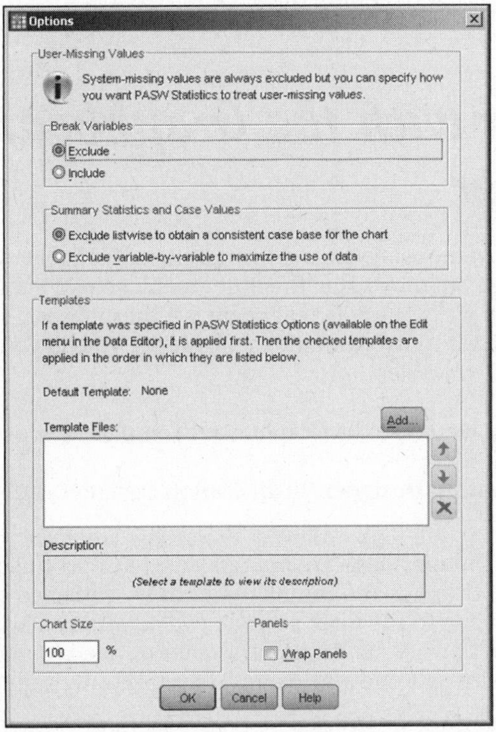

Figure 11-10: Options you can apply to a chart.

When you define the characteristics of a variable, you can specify that certain values be considered missing values. The options in the Break Variables area let you decide whether you want those values included or excluded from your chart. You can also specify how you'd like summary statistics handled. (Missing values are discussed in Chapter 4, and the different types of summary data are described in Chapter 13.)

Templates are files that contain all or part of a chart definition. You can insert one or more names of template files into the list in this window, and SPSS will apply those template definitions as the default starting points for all charts you build. You create a template file from a finished chart displayed in SPSS Statistics Viewer. You find out more on making templates later in this chapter.

Templates come in handy only when you want to build lots of similar charts. You can use the Chart Size option to make the generated charts smaller or larger, as needed.

The Wrap Panels option determines how the panels are displayed when you have a number of them in a chart. SPSS is using the word *panel* to refer to the rectangular area in SPSS Statistics Viewer in which a chart is placed. Normally the panels shrink automatically to fit; if you select this option, however, they remain full-size and wrap to the next line.

Building Graphs with the Graphboard Template Chooser

The charts you build by choosing Graphs⇨Graphboard Template Chooser are similar to those you build by using the other menu selections, but you get less guidance along the way. The way you begin is a bit different. You indicate the variable you want to use, and the menu shows you all your choices for that combination of variables.

There are two big reasons to use the Graphboard Template Chooser:

- ✔ **You own Visualization Designer.** Visualization Designer is discussed a bit in Chapter 22.

- ✔ **You want to make maps.** Maps are not possible in Chart Builder, but they're popular in Graphboard Template Chooser. You need some way to align your data to the map, so you need a variable like State or County. If you don't own Visualization Designer or need to make a map, it's best to stick with just one approach, so we take only a quick peek at this option.

First, you select the variables you want to include (Ctrl+click to select multiple variables), which causes the available kinds of charts (such as bar, dot, or line) to appear onscreen, as shown in Figure 11-11. Using the tabs at the top of the screen (Basic, Detailed, Titles, Options), you can choose screens that allow you to set the options. Note the three "choropleth" options. Those are your maps, but the GSS Survey dataset has broad regions, not more specific geographic info, so mapping is not an option with this data.

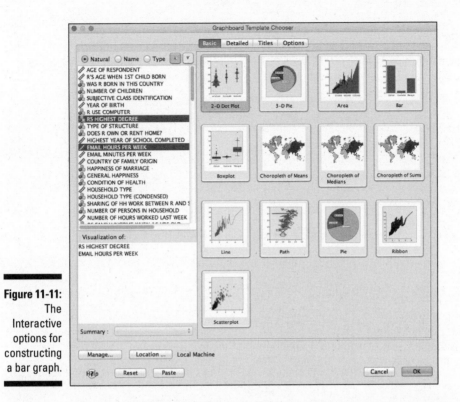

Figure 11-11:
The Interactive options for constructing a bar graph.

If you get things balled up, it's easy to restart. The Reset button removes everything you've entered in all the tabbed windows and restores all the defaults.

The Help button provides some information about whichever list of options is displayed at the moment.

Finding graphs in unexpected places

The Graphs menu isn't the only place that you'll find powerful graphics options. Most analysis methods, for example, will have a Plots tab. These specialty graphics may not be found in the same form, or with the same features, in the Graphs menu. They also tend to be turned off by default, so if you don't request them, you usually won't get them.

Here is a quick list of places to look:

- The Descriptive Statistics submenu of Analyze has techniques that allow you to explore data including Frequencies and Explore.

- The Plots tab for many techniques in the Analyze Menu.

(continued)

(continued)

✔ Dedicated menus for certain specialty modules. For instance, the best graphics for Time Series data are in the Forecasting menu (see the following figure).

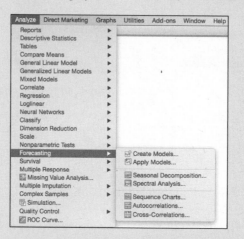

Analysis techniques whose primary purpose is the production of a graphiclike Correspondence Analysis or the brand-new Spatial and Temporal Modeling menu (see the following figure). Decision trees are a highly visual technique as well.

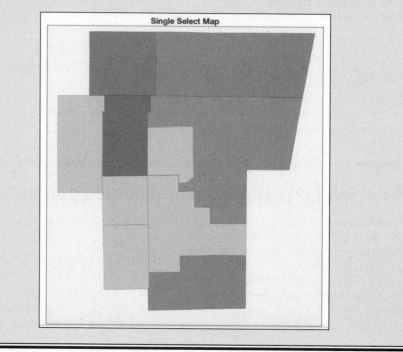

Single Select Map

Chapter 12

Building Graphs Using the Chart Builder

In This Chapter

▶ Generating simple graphs and charts

▶ Creating fancy (multivariable) charts from your data

▶ Displaying scatterplots, histograms, pie charts, and area graphs

*T*he first part of this chapter provides examples of simple graphs and charts and shows you how to build the graphs that you're already familiar with. The second part of this chapter shows examples of graph types that may be less familiar and are more complex. This chapter provides instructions on how to create all the different graph types in a simple, step-by step procedure.

Although the chapter doesn't cover every type of chart, you can use the steps in this chapter to produce some neat graphs. When you get the basic idea of how to build graphs using the Chart Builder, you can explore making some graphs on your own.

You can work through the examples in this chapter to get an overview of building graphs using SPSS — not a bad idea for a beginner — or simply choose the look you want your data to have and follow the steps given here to construct the chart that does the job. Either way, when you get a handle on the basics, you can step through the process again and again, using your data, trying variations until you get charts that appear the way you want them to.

When you use the Chart Builder, it's completely safe to drag and drop any variables you want to see into your graph; if the variable doesn't make sense there, the drop will fail. The Chart Builder in SPSS will tell you what will and won't work.

No matter what you try to do while building a graph, your data will never be hurt.

Simple Graphs

The graphs in this section are a great way to use the Chart Builder to show relationships between one or two variables.

Simple scatterplots

A *scatterplot* is simply an *xy* plot where you don't care about interpolating the values — that is, the points aren't joined with lines. Instead, a disconnected dot appears for each data point. The overall pattern of these scattered dots often exposes a pattern or trend.

The following steps show you how to construct a simple scatterplot:

1. **Choose File⇨Open⇨Data and open the `Cars.sav` file.**

   The file is not in the SPSS installation directory. You have to download it from this book's companion website.

2. **Choose Graphs⇨Chart Builder.**

3. **In the Choose From list, select Scatter/Dot.**

4. **Select the simplest scatterplot diagram (the one with the Simple Scatter tooltip), and drag it to the panel at the top.**

5. **In the Variables list, select Horsepower and drag it to the rectangle labeled X-Axis in the diagram.**

   In a scatterplot, both the *x*-axis and *y*-axis variables are scale. Look for the ruler icon when dragging your variable over.

6. **In the Variables list, select Miles Per Gallon (MPG) and drag it to the rectangle labeled Y-Axis in the diagram.**

7. **Click OK.**

   The chart in Figure 12-1 appears.

Each dot on the scatterplot in Figure 12-1 represents both the horsepower and miles per gallon for each car. The most obvious fact you can derive from this is that the miles per gallon depends largely on the horsepower. In the pattern of the dots, it's easy to see a normal line from the upper left to the lower right. You can see an inverse relationship between horsepower and miles per gallon. As the horsepower of a car increases, there is a decrease in the miles per gallon that the car gets.

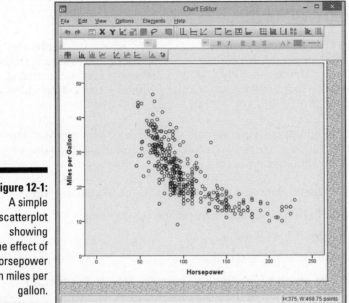

Figure 12-1:
A simple
scatterplot
showing
the effect of
horsepower
on miles per
gallon.

Simple dot plots

No plot is simpler to produce than the *dot plot*. It has only one dimension.
Although SPSS groups it among the scatterplots, there's nothing scattered
about it. It actually presents data more like a bar chart. As you build this type
of chart, you'll notice that you won't need a *y*-axis.

It's easy to create a dot plot. You select the dot plot as the type of graph you
want and then select one variable. SPSS does the rest. The following steps
guide you through the process of creating a simple dot plot:

1. **Choose File⇨Open⇨Data and open the `Employee data.sav` file.**

   The file is in the SPSS installation directory.

2. **Choose Graphs⇨Chart Builder.**

3. **In the Choose From list, select Scatter/Dot.**

4. **Select the sixth graph image (the one with the Simple Dot Plot tooltip)
   and drag it to the panel at the top.**

5. **In the Variables list, select Date of Birth and drag it to the X-Axis
   rectangle.**

6. **Click OK.**

   The chart shown in Figure 12-2 appears.

Simple bar graphs

A *bar graph* is a comparison of relative magnitudes. Simple bar graphs and simple line graphs are the most common ways of charting statistics. It would make an interesting statistical study to determine which is more common. The results could be displayed as either a bar graph or a line graph, whichever is more popular.

A fundamental bar graph is simple enough that the decisions you need to make when preparing one are almost intuitive. The following steps can be used to generate a simple bar graph:

1. **Select File⇨Open⇨Data and open the `Employee data.sav` file.**

   The file is in the SPSS installation directory.

2. **Choose Graphs⇨Chart Builder.**

3. **In the Choose From list, select Bar.**

4. **Select the first graph image (the one with the Simple Bar tooltip) and drag it to the panel at the top of the window.**

5. **In the Variables list, select Employment Category and drag it to the X-Axis rectangle.**

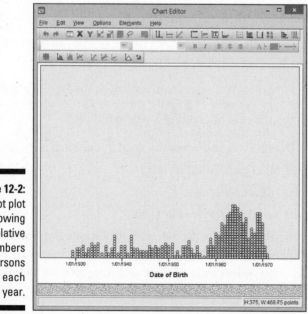

Figure 12-2:
A dot plot showing the relative numbers of persons born in each year.

6. **In the Variables list, select Current Salary and drag it to the Count rectangle.**

    The label changes from Y-Axis to Mean to indicate the type of variable that will now be applied to that axis.

7. **Click OK.**

    The bar graph in Figure 12-3 appears.

The display of data is similar in a line chart and a bar chart. If you decide to display data as a bar chart, you should probably try the same data as a line graph to see which you prefer.

Simple error bars

Some errors come from flat-out mistakes — but those aren't the errors we talk about here. Statistical sampling can help you arrive at a conclusion, but that conclusion has a margin of error. This margin can be calculated and quantified according to the size of the sample and the distribution of the data. For example, suppose you want to know how typical the result is when you calculate the mean of all values for a particular variable — for any one case, the mean could be as big as the largest value or as small as the

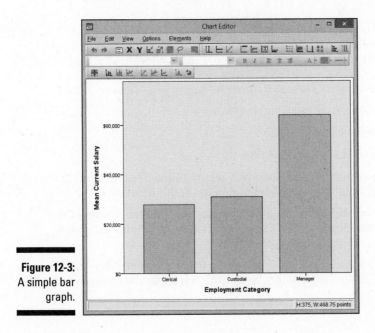

Figure 12-3: A simple bar graph.

smallest. The maximum and minimum are the extremes of the possible error. You can choose values and mark the points that contain, say, 90 percent of all values. Marking these points on graphs creates *error bars*.

You can add error bars to the display of most types of graphs. For example, you could add error bars to the simple bar graph presented earlier (refer to Figure 12-3) by making selections in the Element Properties dialog box. If you've worked through any of the examples, you'll know Element Properties as that pesky window that pops up every time you construct a chart.

For an example of adding error bars to a bar chart, follow the same procedure described previously in the "Simple bar graphs" section — but just before the final step (clicking OK to produce the chart), do the following:

1. **If the Element Properties window is not displayed, click the Element Properties button to put that window onscreen.**

2. **In the Element Properties window, make sure that a check mark appears in the Display Error Bars option and that Bar1 is selected.**

3. **Select Confidence Intervals and set its value to 95%.**

4. **Click Apply.**

5. **Close the Element Properties window by clicking the Close button.**

6. **Click OK.**

    The chart shown in Figure 12-4 appears.

Another way to display the same data is to display the range of errors without displaying the full range of all values. To do so, follow these steps:

1. **Choose File⇨Open⇨Data and open the `Employee data.sav` file.**

    The file is in the SPSS installation directory.

2. **Choose Graphs⇨Chart Builder.**

3. **In the Choose From list, select Bar.**

4. **Select the seventh graph image (the one with the Simple Error Bar tooltip) and drag it to the panel at the top of the window.**

5. **In the Element Properties window, make sure that the Display Error Bars option is checked, Bar1 is selected, the Confidence Intervals option is selected, and the Level is set to 95%.**

    Any time you change a setting or a value in the Element Properties dialog box, you must click the Apply button to have the change reflected in your chart.

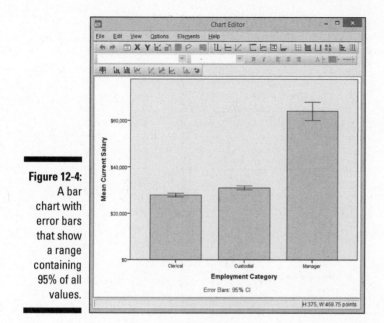

Figure 12-4:
A bar
chart with
error bars
that show
a range
containing
95% of all
values.

6. **In the Variables list, select Employment Category and drag it to the X-Axis rectangle.**

7. **In the Variables list, select Current Salary and drag it to the Mean rectangle.**

   The label changes from Y-Axis to Mean to indicate the type of data that will be displayed on that axis.

8. **Click OK.**

   The bar graph in Figure 12-5 appears.

This example displays the result of one way of making error calculations: The magnitude of the error is based on 95% of all values being within the upper and lower error bounds. If you prefer, you can base the error on the bell curve and mark the upper and lower errors at some multiple of the standard error or standard deviation.

Simple histograms

A *histogram* represents the number of items that appear within a range of values. You can use a histogram to look at a graphic representation of the frequency distribution of the values of a variable. Histograms are useful for

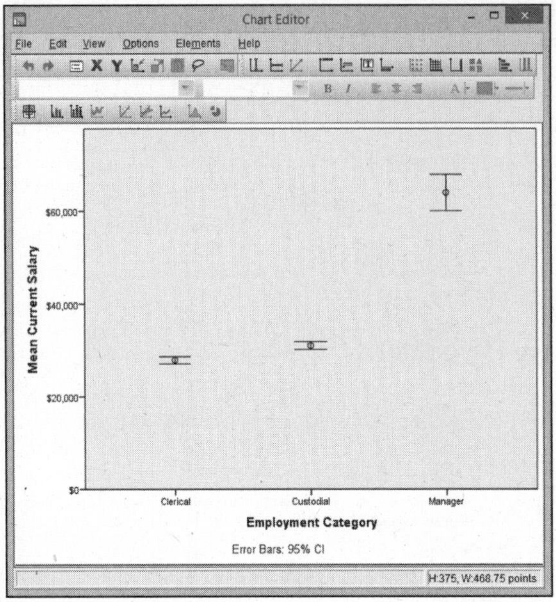

Figure 12-5:
An error bar graph, showing the mean values as dots and the upper and lower bounds of the error.

demonstrating the patterns in your data when you want to display information to others rather than discover data patterns for yourself.

You can use the following steps to create a simple histogram that displays the frequency of the ages at which the respondent's first child was born:

1. **Choose File⇨Open⇨Data and open the GSS2012.sav file.**

   The file is not in the SPSS Statistics installation directory. You have to download it from this book's companion website.

2. **Choose Graphs⇨Chart Builder.**

   The Chart Builder dialog box appears.

3. **In the Choose From list, select Histogram.**

4. **Drag the first graph diagram (the one with Simple Histogram tooltip) to the panel at the top of the window.**

5. **In the Variables list, select the R'S AGE WHEN 1ST CHILD BORN variable and drag into the X-Axis rectangle in the panel.**

6. **Click OK.**

   The histogram shown in Figure 12-6 appears.

The graph in Figure 12-6 looks like a bar chart, but it isn't. The height of each bar does not represent the mean or an average — the height is determined by the largest value.

The meaning of a graph of this sort is not intuitive; you may want to add a note explaining what it means.

Population pyramids

A *population pyramid* provides an immediate comparison of the number of items that fall into categories. It's called a pyramid because it often takes a triangular shape — wide at the bottom and tapering to a point at the top. The following steps can be followed to build an example pyramid histogram chart:

1. **Choose File⇨Open⇨Data and open the GSS2012.sav file.**

    The file is not in the SPSS installation directory. You have to download it from this book's companion website.

2. **Choose Graphs⇨Chart Builder.**

3. **In the Choose From list, select Histogram.**

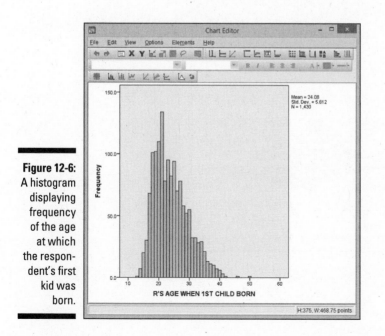

Figure 12-6: A histogram displaying frequency of the age at which the respondent's first kid was born.

4. Drag the fourth graph diagram (the one with the Population Pyramid tooltip) to the panel at the top of the window.

5. In the Variables list, do the following:

    a. Select the Respondent's Sex variable and drag it to the Split Variable rectangle.

    This is a categorical variable with two possible values, so one category will be placed on each side of the center line.

    b. Select R'S AGE WHEN 1ST CHILD BORN and drag it to the Distribution Variable rectangle.

6. Click OK.

    The chart shown in Figure 12-7 appears.

You can create pyramid histograms based on categorical variables with three, four, or more values. The plot produced will consist of as many pairs as needed (and even a single-sided pyramid for one category, if necessary) to display bars that show the relative number of occurrences of different values in the categories.

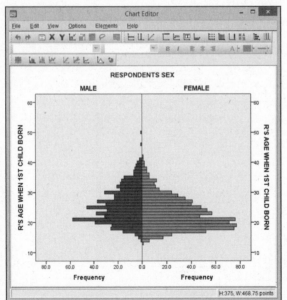

Figure 12-7:
A population pyramid shows the occurrence of values within categories.

Stacked area charts

An *area graph* is really a line graph, or a collection of line graphs, with the areas below the lines filled in to represent the mean of one or more values at the various points.

A *stacked area chart* is a chart with more than one variable being displayed along the *x*-axis. The values are stacked in such a way that the ups and downs of the lower value in the chart affect the upper values in the chart. That is, the chart is not a group of independent lines; instead, it represents a cumulative total — to which each variable displayed adds a value.

Step 5c of the following procedure can be repeated for the inclusion of two or more variables. They all appear in the legend at the upper right, and each variable provides the value for one layer of the stack.

If you include more than one variable, make sure that the variables you select for stacking have similar ranges of values so the scale on the left side will make sense for all of them. If, for example, one variable ranges into the thousands and the other doesn't go over a hundred, the smaller one will compress itself visually — and come out in the final graph as a line.

When you select multiple variables for stacking, be sure to select them in the order you want them stacked; the first one you select will remain on top. The second one you select will be placed under it, and so on.

The two types of area charts — simple and stacked — act the same when you construct them. You can select the stacked chart and produce a simple area chart, or you can start with the simple area chart and stack your variables.

Follow these steps to produce a stacked area chart with two stacked variables:

1. **Choose File⇨Open⇨Data and open the `Employee data.sav` file.**

   The file is in the SPSS installation directory.

2. **Choose Graphs⇨Chart Builder.**

3. **In the Choose From list, select Area.**

4. **Drag the second graph diagram (the one with the Stacked Area tooltip) to the panel at the top of the window.**

5. In the Variables list, do the following:

a. Select the Educational Level variable and drag it to the X-Axis rectangle.

b. Select Current Salary and drag it to the Y-Axis rectangle.

c. Select Beginning Salary and drag it to the Current Salary rectangle. A plus sign appears when you drag over it.

Be sure to drag it to the plus sign and not simply to the rectangle in general. (The plus sign appears at the top of the rectangle when you drag the new variable's name across it.)

The Create Summary Group dialog box appears. After dragging in Beginning Salary, the Stacked: Set Color rectangle is relabeled "Index."

6. Click OK.

The area chart shown in Figure 12-8 appears.

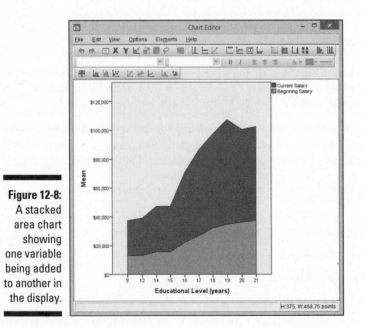

Figure 12-8:
A stacked area chart showing one variable being added to another in the display.

Fancy Graphs

Fancy graphs are just like they sound: fancy. Whether you're trying to show relationships with lines, bars, dots, or more, you can use the Chart Builder in SPSS to create these graphs to show relationships between more than two variables.

Charts with multiple lines

A *line chart* works well as a visual summary of categorical values. Line charts are also useful for displaying timelines because they demonstrate up and down trends so well. Line graphs are popular because they're easy to read. If they're not *the* most common type of statistical chart, they're a contender for the title.

You can have more than one line appear on a chart by adding more than one variable name to an axis. But the variables must contain a similar range of values before they can be represented by the same axis. For example, if one variable ranges from 0 to 1,000 pounds and another variable ranges from 1 to 2 pounds, the values of the second variable will show up as a straight line, regardless of how much it actually fluctuates.

The following steps generate a basic multiline graph:

1. **Choose File⇨Open⇨Data and open the `Cars.sav` file.**

   The file is not in the SPSS installation directory. You have to download it from this book's companion website.

2. **Choose Graphs⇨Chart Builder.**

3. **In the Choose From list, select Line to specify the general type of graph to be constructed.**

4. **To specify that this graph should contain multiple lines, select the second diagram (the one with the Multiple Line tooltip) and drag it to the panel at the top.**

   The Element Properties dialog box appears, but you can close it because the default values work fine.

5. **In the Variables list, right-click Model Year and select Ordinal. Then select Model Year and drag it to the rectangle named X-Axis in the diagram.**

6. **In the Variables list, select Engine Displacement and drag it to the Y-Axis rectangle in the panel at the top where it says Count.**

   The word *Mean* is added to the annotation because the values displayed on this axis will be the mean values of the engine displacement.

7. **In the Variables list, select Country of Origin and drag it to the Set Color rectangle in the panel at the top.**

8. **Click OK.**

   The chart shown in Figure 12-9 appears.

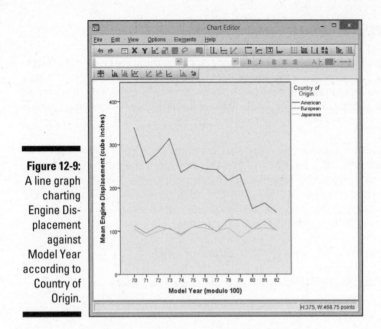

Figure 12-9:
A line graph
charting
Engine Dis-
placement
against
Model Year
according to
Country of
Origin.

The following steps generate a multiline graph with two variables in the Y-axis:

1. **Choose File⇨Open⇨Data and open the `Cars.sav` file.**

   The file is not in the SPSS installation directory. You have to download it from this book's companion website.

2. **Choose Graphs⇨Chart Builder.**

3. **In the Choose From list, select Line to specify the general type of graph to be constructed.**

4. **To specify that this graph should contain multiple lines, select the second diagram (the one with the Multiple Line tooltip) and drag it to the panel at the top.**

   The Element Properties dialog box appears, but you can close it because the default values work fine.

5. **In the Variables list, right-click Model Year and select Ordinal. Then select Model Year and drag it to the rectangle named X-Axis in the diagram.**

6. **In the Variables list, select Engine Displacement and drag it to the Y-Axis rectangle in the panel at the top where it says Count.**

   The word *Mean* is added to the annotation because the values displayed on this axis will be the mean values of the engine displacement.

7. In the Variables list, select Horsepower and drag it to the Y-axis rectangle in the panel at the top.

Be careful how you drop Horsepower. To add Horsepower as a new variable, you want to drop it on the little box containing the plus sign, as shown in Figure 12-10. If you drop the new name on top of the one that's already there, the original variable could be replaced.

Figure 12-10: Adding another variable to the Y-axis.

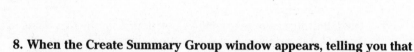

8. When the Create Summary Group window appears, telling you that SPSS is combining the two variables along the Y-axis, click OK.

9. Click OK.

The chart shown in Figure 12-11 appears.

Figure 12-11: A line graph with two variables in the Y-axis.

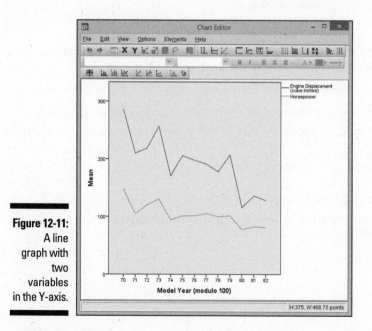

The variables you choose as members of the Y-axis must have a similar range of values to make sense. For example, if you were to choose Vehicle Weight and Engine Displacement to be charted together, the result wouldn't be all that interesting; the Vehicle Weight values would be in the thousands and the Engine displacement, regardless of their variation, would all appear in a single line at the bottom. In Chapter 18, we discuss how to edit these graphs to display information from the graphs in a more impactful way.

Colored scatterplots

Colored scatterplots can be dramatic in appearance, but clarity is not their strongest point. Because the scatterplot is drawn on a two-dimensional surface, you may find it difficult to envision where each point is supposed to appear in space. On the other hand, if your data distributes appropriately on the display, the resulting chart may demonstrate the concept you're trying to get across.

The following example uses the same data as in the preceding example but displays it in a different way, as a three-dimensional plot:

1. **Choose File⇨Open⇨Data and open the `Cars.sav` file.**

   The file is not in the SPSS Statistics installation directory. You have to download it from this book's companion website.

2. **Choose Graphs⇨Chart Builder.**

3. **In the Choose From list, select Scatter/Dot.**

4. **Select the second scatterplot diagram (the one with the Grouped Scatter tooltip) and drag it to the panel at the top.**

5. **In the Variables list, do the following:**

   a. Select Engine Displacement and drag it to the X-Axis rectangle.

   b. Select Miles Per Gallon and drag it to the Y-Axis rectangle.

   c. Select Country of Origin and drag it to the Set Color rectangle.

6. **Click OK.**

   The graph shown in Figure 12-12 appears.

   We recommend using a bubble chart for displaying three-scale variables over a 3D scatterplot.

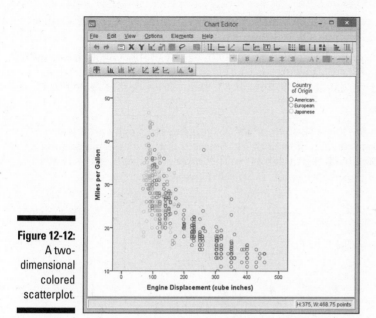

Figure 12-12:
A two-dimensional colored scatterplot.

Scatterplot matrices

A *scatterplot matrix* is a group of scatterplots combined into a single graphic image. You choose a number of scale variables and include them as a member of your matrix, and SPSS creates a scatterplot for each possible pair of variables. You can make the matrix as large as you like — its size is controlled by the number of variables you include.

The following steps walk you through the creation of a matrix:

1. **Choose File⇨Open⇨Data and open the `Cars.sav` file.**

   The file is not in the SPSS installation directory. You have to download it from this book's companion website.

2. **Choose Graphs⇨Chart Builder.**

3. **In the Choose From list, select Scatter/Dot.**

4. **Select the seventh graph image (the one with the Scatterplot Matrix tooltip) and drag it to the panel at the top.**

5. **In the Variables list, drag Engine Displacement to the Scatterplot Matrix rectangle in the panel at the top.**

   The selected name replaces the label in the rectangle.

6. **Drag the variable names Miles Per Gallon, Horsepower, and Vehicle Weight to the plus sign on the *x*-axis.**

   The labels may or may not change with each variable you add, depending on their length and the amount of space available. All your labels appear in the list at the bottom of the Element Properties dialog box.

7. **Click OK.**

   The chart in Figure 12-13 appears. As you can see, each variable is plotted against each of the others. Notice that the scatterplots along the diagonal from the upper left to the lower right are blank — that's because it's useless to plot a variable against itself.

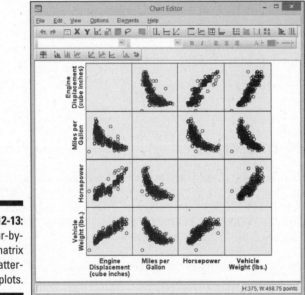

Figure 12-13:
A four-by-four matrix of scatter-plots.

Stacked bar charts

A *stacked bar chart* is similar to the clustered bar chart in that it displays multiple values of a variable for each value of a categorical variable. But it does so by stacking them instead of placing them side by side. The following chart displays the same data as the preceding example, but it emphasizes different aspects of the data.

Follow these steps to create a stacked bar chart:

1. **Choose File⇨Open⇨Data and open the `Cars.sav` file.**

   The file is not in the SPSS installation directory. You have to download it from this book's companion website.

2. **Choose Graphs⇨Chart Builder.**

3. **In the Choose From list, select Bar.**

4. **Select the third graph image (the one with the Stacked Bar tooltip) and drag it to the panel at the top of the window.**

5. **In the Variables list, do the following:**

   a. *Right-click Model Year and select Ordinal.*

   b. *Select Model Year and drag it to the X-Axis rectangle.*

   c. *Select Miles Per Gallon and drag it to the Count rectangle.*

      The rectangle was originally labeled Y-Axis. The label changed to help you understand the type of variable that needs to be placed there.

   d. *Select Country of Origin and drag it to the rectangle in the upper-right corner, the one now labeled Stack Set Color.*

6. **Click OK.**

   The graph in Figure 12-14 appears.

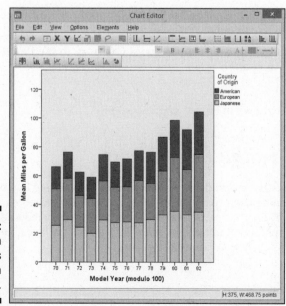

Figure 12-14:
A bar graph with values displayed in stacks.

Pie charts

Pie charts are the easiest kind to spot — they're the only charts that show up as circles. The purpose of a *pie chart* is simply to show how something (the "whole") is divided into pieces — whether two, ten, or any other number. Each slice in the pie chart represents its percentage of the whole. For example, if a slice takes up 40% of the total pie, that slice represents 40% of the total number. A pie chart is also called a *polar chart,* so SPSS calls this option Pie/Polar.

In the following steps, you construct a pie chart:

1. **Choose File⇨Open⇨Data and open the Bank.sav file.**

   The file is not in the SPSS installation directory. You have to download it from this book's companion website.

2. **Choose Graphs⇨Chart Builder.**

3. **In the Choose From list, select Pie/Polar.**

4. **Drag the pie diagram to the panel at the top of the window.**

5. **Click the Groups/Point ID tab.**

6. **Click the Columns Panel variable.**

   A rectangle with Panel should appear in the summary area.

7. **Click the Options button and select Wrap Panels at the bottom. Then click OK.**

8. **In the Variables list, drag Sex of Employee to the Slice By rectangle at the bottom of the panel.**

9. **In the Variables list, drag Employment Category to the Panel rectangle at the top of the panel.**

10. **Click OK.**

    The pie chart shown in Figure 12-15 appears.

Clustered range bar graphs

The *clustered range bar graph* displays the relationship among five variables. No other chart can be used to so clearly display so many variables. This example demonstrates the relationships among five employee variables: years of employment, income ranges, years with employer, ages, and the years they've lived at their current addresses.

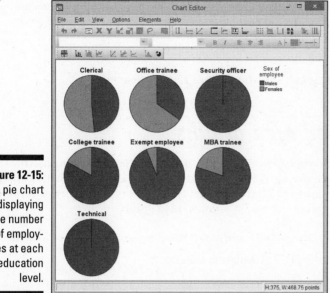

Figure 12-15:
A pie chart displaying the number of employees at each education level.

Do the following to build a clustered range bar graph:

1. **Choose File⇨Open⇨Data and open the file named `Customer_dbase. sav`.**

   The file is in the SPSS installation directory.

2. **Choose Graphs⇨Chart Builder.**

3. **In the Choose From list, select High-Low.**

4. **Drag the third graph diagram (the one with the Clustered Range Bar tooltip) to the panel at the top of the window.**

5. **In the Variables list, do the following:**

   a. *Drag the category Years With Current Employer[empcat] variable to the Cluster on X rectangle in the upper-right corner.*

   b. *Drag the Income Category in Thousands variable to the X-Axis rectangle at bottom.*

   c. *Right-click Age In Years and make it Scale. Then drag the Age In Years variable to the High Variable rectangle at the top left.*

   d. *Right-click Years with Current Employer [employ] and make it Scale. Then drag the measure Years with Current Employer [employ] variable to the Low Variable rectangle at the center left.*

e. *Right-click Years at Current Address (address) and make it Scale. Then drag the Years at Current Address (address) variable to the Close Variable rectangle at the bottom left.*

6. Click OK.

The high-low graph shown in Figure 12-16 appears.

Differenced area graphs

A *differenced area graph* provides a pair of line graphs that emphasize the difference between two variables, filling the area between them with a solid color. The two graphs are plotted against the points of a categorical variable. The following steps produce a differenced area graph:

1. Choose File⇨Open⇨Data and open the `property_assess.sav` file.

The file is in the SPSS installation directory.

2. Choose Graphs⇨Chart Builder.

3. In the Choose From list, select High-Low.

4. Drag the fourth graph diagram (the one with the Differenced Area tooltip) to the panel at the top of the window.

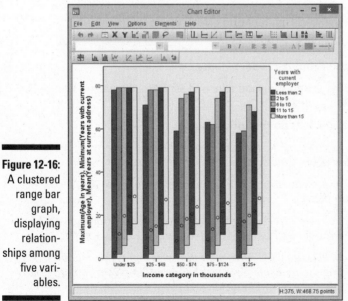

Figure 12-16: A clustered range bar graph, displaying relationships among five variables.

5. In the Variables list, do the following:

a. *Drag the Township variable to the X-Axis rectangle.*

b. *Drag the Sale Value of House variable to either of the Y-Axis rectangles.*

c. *Drag the Value at Last Appraisal variable to the other Y-Axis rectangle.*

6. Click OK.

The differenced area chart shown in Figure 12-17 appears.

Dual-axis graph

Many of the graphic forms allow you to plot two or more variables on the same chart, but they must always be plotted against the same scale. In the *dual-axis graph,* two variables are plotted and two different scales are used to plot them. As a result, the values don't require the same ranges (as they do in the other plots); the curves and trends of the two variables can be easily compared, even though they're on different scales.

Two variables with different ranges that vary across the same set of categories can be plotted together, as shown in the following example:

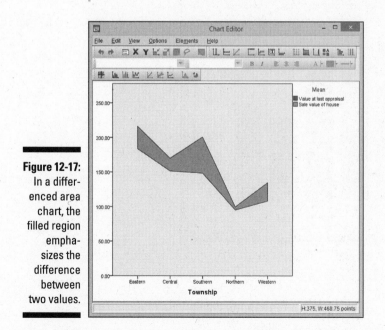

Figure 12-17: In a differenced area chart, the filled region emphasizes the difference between two values.

1. **Choose File➪Open➪Data and open the** property_assess.sav **file.**

   The file is in the SPSS installation directory.

2. **Choose Graphs➪Chart Builder.**

3. **In the Choose From list, select Dual Axes.**

4. **Drag the first diagram (the one with the Dual Y Axes with Categorical X Axis tooltip) to the panel at the top of the window.**

5. **In the Variables list, do the following:**

   a. Drag the Value at Last Appraisal variable to the Y-Axis rectangle on the left.

   b. Drag the Sale Value of House variable to the Y-Axis rectangle on the right, which is now named Count.

   c. Drag the Township variable to the X-Axis rectangle.

6. **Click OK.**

   The dual-axis graph shown in Figure 12-18 appears.

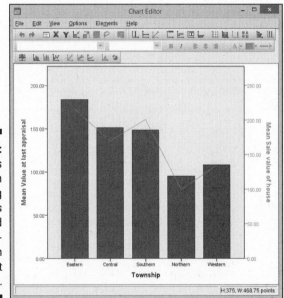

Figure 12-18: A dual-axis graph displaying the curves inscribed by two variables with different ranges.

Part V
Analyzing Data

Get to know the Paired-Samples T Test in an article at www.dummies.com/extras/spss.

In this part . . .

- ✔ Navigate the huge Analysis menu with confidence.
- ✔ Identify which test you need and why.
- ✔ Understand the output so that you can explain it to others.

Chapter 13

Using Descriptive Statistics

· ·

· ·

Summaries of individual variables provide the basis for more complex analysis (as you see in the next few chapters). They also help establish base rates, answer important questions (for example, the percent of satisfied customers), allow users to check sample size and the data for unusual cases or errors, and provide insights into ways in which you may combine different groups. Ideally, you want to obtain as much information as possible from your data. In practice, however, given the measurement level of the variables, only some information is meaningful.

In this chapter, we begin by discussing level of measurement. Next, we run the frequencies procedure to obtain summary statistics for both categorical and continuous variables. Finally, we use the descriptives procedure to summarize continuous variables.

Looking at Levels of Measurement

The level of measurement of a variable determines the appropriate statistics, and graphs that can be used to describe the data. For example, if we have a variable like marital status, it wouldn't make sense to ask for the mean of this variable; instead, we may ask for the percentages associated with the different categories. In addition, level of measurement also determines the kind of research questions we can answer, so it's a critical step in the research process.

The term *levels of measurement* refers to the coding scheme or the meaning of the numbers associated with each variable. Many statistical techniques are appropriate only for data measured at particular levels or combinations of levels. Different statistical measures are appropriate for different types of variables, and the statistical summaries depend on the level of measurement.

Defining the four levels of measurement

Introductory statistics textbooks present four levels of measurement, each defined by certain properties. Each successive level can be said to contain the properties of the preceding types and records information at a higher level. The four levels of measurement are as follows:

- **Nominal:** For nominal data, each value represents a category. There is no inherent order to the categories. For example, the variable gender may be coded as 0 (male) and 1 (female), but all these values tell us is that we have two distinct categories, *not* that one category has more or less or is better or worse than the other.

- **Ordinal:** For ordinal data, each value is a category, but there is a meaningful order or rank to the categories. However with ordinal data, there is not a measurable distance between categories. For example, if we're measuring the outcome of a foot race, we can determine which contestant came in first, second, third, and so on. However, based on the ranking, we can't tell how much faster each competitor was compared to the others, nor can we say that the difference between first and second place is the same as the difference between second and third place. Other examples of ordinal variables are attitudinal questions with categories, such as Strongly Disagree (1), Disagree (2), Neutral (3), Agree (4), and Strongly Agree (5), or variables such as income coded into categories representing ranges of values.

- **Interval:** For interval data, a one-unit change in numeric value represents the same change in quantity regardless of where it occurs on the scale. For example, for a variable like temperature measured in Fahrenheit, the difference between 20 degrees and 21 degrees (1 unit) is equal to the difference between 50 degrees and 51 degrees. In other words, they have equal intervals between points on the scale.

- **Ratio:** For ratio data, you have all the properties of interval variables with the addition of a true zero point, representing the complete absence of the property being measured. For example, temperature measured in Fahrenheit is measured on an interval scale, because zero degrees does not represent the absence of temperature. However, a variable like number of purchases is a ratio variable because zero indicates no purchases. Ratios can then be calculated (for example, eight purchases represents twice as many purchases as four purchases).

These four levels of measurement are often combined into two main types:

- ✔ **Categorical:** Nominal and ordinal measurement levels
- ✔ **Continuous (or scale):** Interval and ratio measurement levels

Defining summary statistics

The most common way to summarize variables is to use measures of central tendency and variability:

- ✔ **Central tendency:** One number that is often used to summarize the distribution of a variable. Typically, we think of central tendency as referring to the "average" value. There are three main measures of central tendency:

  - **Mode:** The category or value that contains the most cases. This measure is typically used on nominal or ordinal data and can easily be determined by examining a frequency table.

  - **Median:** The midpoint of a distribution; it is the 50th percentile. If all the cases for a variable are arranged in order according to their value, the median is the value that splits the data into two equally sized groups.

  - **Mean:** The mathematical average of all the values in the distribution (that is, the sum of the values of all cases divided by the total number of cases).

- ✔ **Variability:** The amount of spread or dispersion around the measure of central tendency. There are a number of measures of variability:

  - **Maximum:** The highest value for a variable.

  - **Minimum:** The lowest value in the distribution.

  - **Range:** The difference between the maximum and minimum values.

  - **Variance:** Provides information about the amount of spread around the mean value. It's an overall measure of how clustered data values are around the mean. The variance is calculated by summing the square of the difference between each value and the mean and dividing this quantity by the number of cases minus one. In general terms, the larger the variance, the more spread there is in the data; the smaller the variance, the more the data values are clustered around the mean.

  - **Standard deviation:** The square root of the variance. The variance measure is expressed in the units of the variable squared. This can cause difficulty in interpretation, so more often, the standard deviation is used. The standard deviation restores the value of variability to the units of measurement of the original variable.

We care about level of measurement because it determines appropriate summary statistics and graphs to describe the data. Table 13-1 summarizes the most common summary statistics and graphs for each of the measurement levels used by SPSS.

Table 13-1	**Level of Measurement and Descriptive Statistics**		
	Nominal	*Ordinal*	*Scale*
Definition	Unordered categories	Ordered categories	Numeric values
Examples	Gender, geographic location, job category	Satisfaction ratings, income groups, ranking of preferences	Number of purchases, cholesterol level, age
Measures of central tendency	Mode	Mode, median	Mode, median, mean
Measures of dispersion	None	Min/max/range	Min/max/range, standard deviation/variance
Graph	Pie or bar	Pie or bar	Histogram

Focusing on Frequencies for Categorical Variables

The most common technique for describing categorical data — nominal and ordinal levels of measurement — is to request a frequency table, which provides a summary showing the number and percentage of cases falling into each category of a variable. Users can also request additional summary statistics like the mode or median, among others.

Here's how to run the frequencies procedure so you can create a frequency table that will allow you to obtain summary statistics for categorical variables:

1. **From the main menu, choose File ⇨ Open ⇨ Data and load the merchandise.sav data file.**

   The file is not in the SPSS installation directory. You have to download it from this book's companion website.

   It contains the customer's purchase history and has 16 variables and 3,338 cases.

2. Choose Analyze ⇨ Descriptive Statistics ⇨ Frequencies.

The Frequencies dialog box, shown in Figure 13-1, appears.

In this example, we want to study the distribution of the variables Payment_Method (Auto Pay, Check, or Credit Card), Premier (Yes or No), and Status (Current or Churned). You can place these variables in the Variable(s) box and each will be analyzed separately.

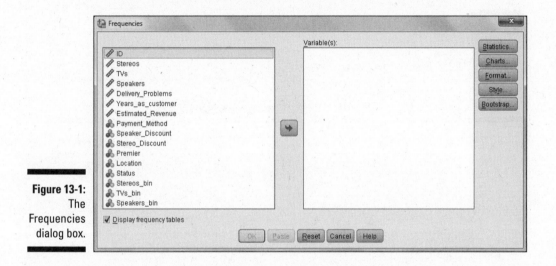

Figure 13-1: The Frequencies dialog box.

3. Select the variables Payment_Method, Premier, and Status, and place them in the Variable(s) box, as shown in Figure 13-2.

If you were to run the Frequencies procedure now, you would get three tables, each showing the distribution of one variable. It's customary though to request additional summary statistics.

4. Click the Statistics button.

The Frequencies: Statistics dialog box, shown in Figure 13-3, appears.

This dialog box provides many statistics, but it's critical that you request only those appropriate for the level of measurement of the variables you placed in the Variable(s) box. For nominal variables, only the mode is suitable.

5. In the Central Tendency section, select the Mode check box, as shown in Figure 13-4.

6. Click Continue.

Requesting a graph, so you can have a visual display of the data, can be useful. That's what we'll do now.

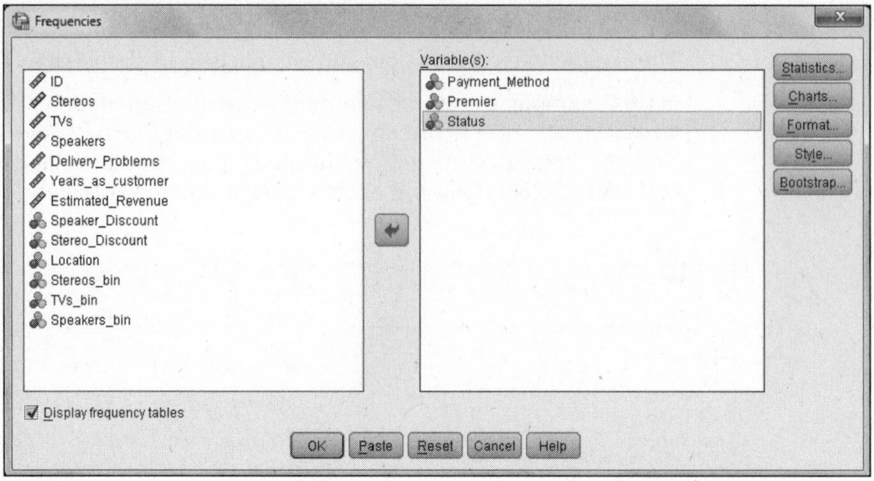

Figure 13-2:
Place the
variables
in the
Variable(s)
box.

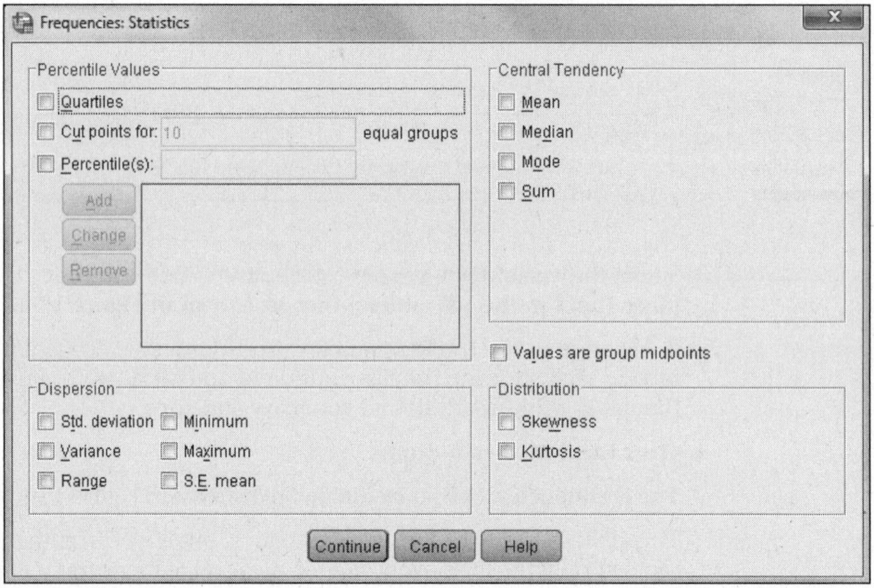

Figure 13-3:
The
Frequen-
cies:
Statistics
dialog box.

7. Click the Charts button.

The Frequencies: Charts dialog box, shown in Figure 13-5, appears.

This dialog box has options for pie charts and bar charts. Either type of chart is acceptable for a nominal variable. Charts can be built using either counts or percentages, but normally percentages are a better choice.

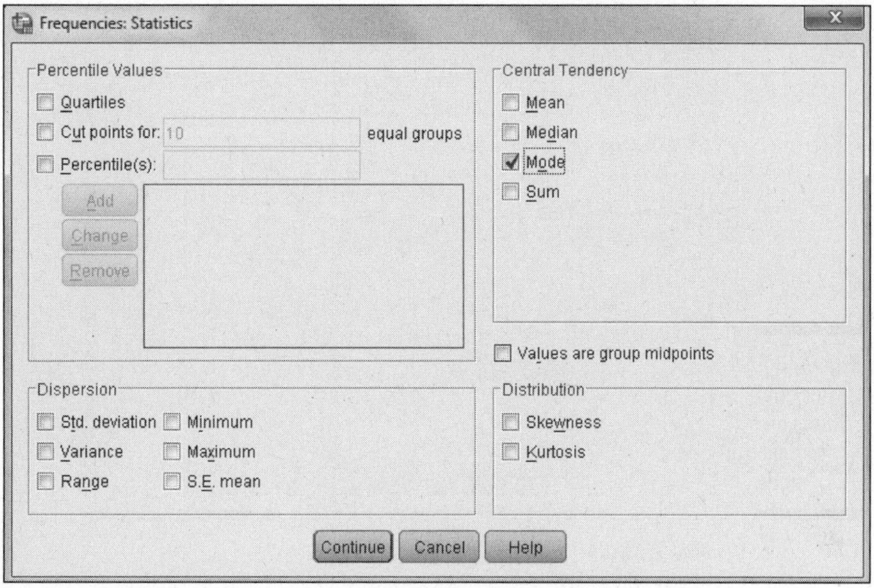

Figure 13-4:
Select the
Mode check
box.

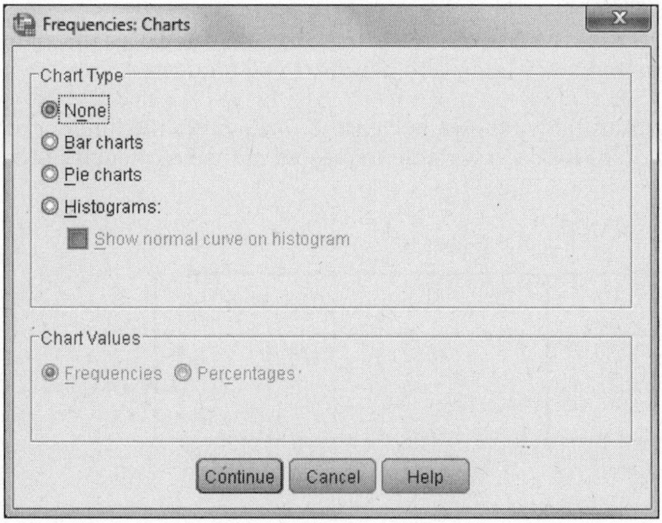

Figure 13-5:
The
Frequen-
cies: Charts
dialog box.

8. **In the Chart Type section, click the Bar Charts radio button;
   in the Chart Values section, click the Percentages radio button
   (see Figure 13-6).**

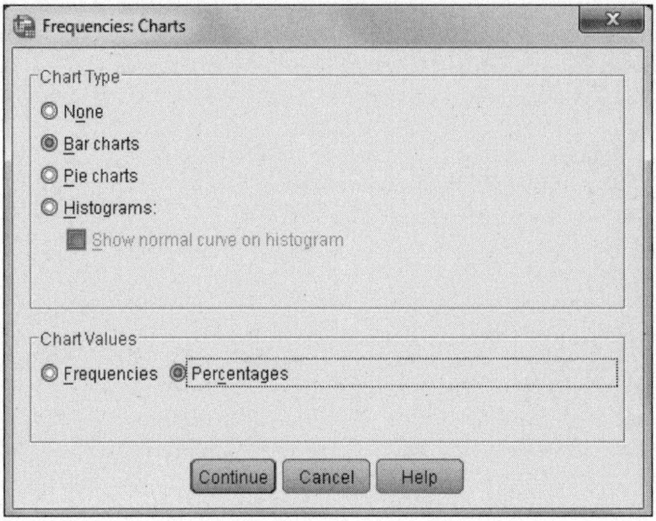

Frequencies: Charts

Chart Type
- ○ None
- ◉ Bar charts
- ○ Pie charts
- ○ Histograms:
  - ☐ Show normal curve on histogram

Chart Values
- ○ Frequencies ◉ Percentages

[Continue] [Cancel] [Help]

Figure 13-6: Click Bar Charts and Percentages.

9. **Click Continue.**

10. **Click OK.**

SPSS runs the frequencies procedure and calculates the summary statistics, frequency table, and bar chart you requested.

The Statistics table (shown in Figure 13-7) displays the number of valid and missing cases for each variable requested in the Frequencies procedure.

Statistics

		Payment_Method	Premier	Status
N	Valid	3338	3338	3338
	Missing	0	0	0
Mode		3	1	2

Figure 13-7: The Statistics table.

Be sure to review this table to check the number of missing cases. In this example, we have 3,338 valid cases and we don't have any missing data.

The Statistics table also displays any additional statistics that were requested. In our case, we asked only for the mode, the category that has the highest frequency, so only the mode is shown for each of the variables. In this example, the mode is represented by values of 3, 1, and 2, respectively, and represents the category of "Credit Card" for Payment_Method, "No" for Premier, and the "Current" group for Status.

We could've asked for additional summary statistics like the mean, and the frequencies procedure would've produced it. This is why it's important to understand measurement level and what statistics are relevant.

The Frequency table (shown in Figure 13-8) shows the distribution of the variable Payment_Method (in this case, we focus on the variable Payment_Method because all the other Frequency tables will have similar information). The information in the Frequency table is comprised of counts and percentages. The Frequency column contains counts, or the number of occurrences of each data value. So, for the variable Payment_Method, it's easy to see why the category "Credit Card" was the mode — 1,926 customers made purchases this way. The Percent column shows the percentage of cases in each category relative to the number of cases in the entire dataset, including those with missing values. In our current example, those 1,926 customers who paid via credit card account for 57.7% of all customers. The Valid Percent column contains the percentage of cases in each category relative to the number of valid (nonmissing) cases. Because there is no missing data, the percentages in the Percent column and in the Valid Percent column are identical. The Cumulative Percent column contains the percentage of cases whose values are less than or equal to the indicated value. Cumulative percent is useful only for variables that are ordinal.

Figure 13-8:
The Frequency table for the Payment_ Method variable.

Payment_Method

		Frequency	Percent	Valid Percent	Cumulative Percent
Valid	Auto Pay	669	20.0	20.0	20.0
	Check	743	22.3	22.3	42.3
	Credit Card	1926	57.7	57.7	100.0
	Total	3338	100.0	100.0	

Bar charts (like the one in Figure 13-9) summarize the distribution that was observed in the Frequency table and allow you to see the distribution. For the variable Payment_Method, more than half of the people fall into the Credit Card category.

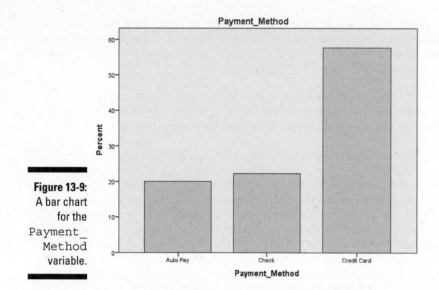

Figure 13-9:
A bar chart
for the
`Payment_`
`Method`
variable.

Understanding Frequencies for Continuous Variables

As we have seen, frequency tables show counts and percentages, which are extremely useful when working with categorical variables. However, for continuous variables that have many values, frequency tables become less useful. For example, if we were working with a variable like income, it wouldn't be very useful to know that there was only one person in the dataset who made $22,222 last year. In this case, it's likely that each response would have a different value, so the frequency table would be very large and not particularly useful as a summary of the variable.

Instead, if the variables of interest are continuous, the frequencies procedure can be useful because of the summary statistics it can produce. To run the frequencies for continuous variables, follow these steps:

1. **From the main menu, choose File ⇨ Open ⇨ Data and load the `merchandise.sav` data file.**

   The file is not in the SPSS installation directory. You have to download it from this book's companion website.

2. **Choose Analyze ⇨ Descriptive Statistics ⇨ Frequencies.**

3. **Select the variables** Stereos, TVs, Speakers, Delivery_Problems, Years_as_customer, **and** Estimated_Revenue, **and place them in the Variable(s) box.**

4. **Deselect the Display Frequency Tables check box, as shown in Figure 13-10.**

   A warning dialog box appears saying, "You have turned off all output. Unless you select any Output Options this procedure will not be run." We receive this warning because at the moment nothing is selected. This is okay because we will now select the summary statistics we want to display.

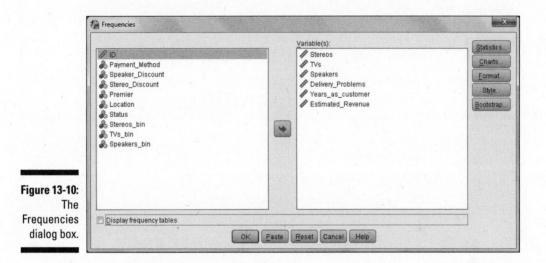

Figure 13-10:
The
Frequencies
dialog box.

5. **Click the Statistics button.**

   The Frequencies: Statistics dialog box appears.

   Several summary statistics are appropriate for scale variables. The statistics can be divided into those summarizing the central tendency, those measuring the amount of variation (dispersion) in the data, different percentile values you can request, and those statistics assessing the shape of the distribution.

6. **In the Central Tendency section, select the Mean, Median, and Mode check boxes; in the Dispersion section, select the Std. Deviation, Minimum, and Maximum check boxes (see Figure 13-11).**

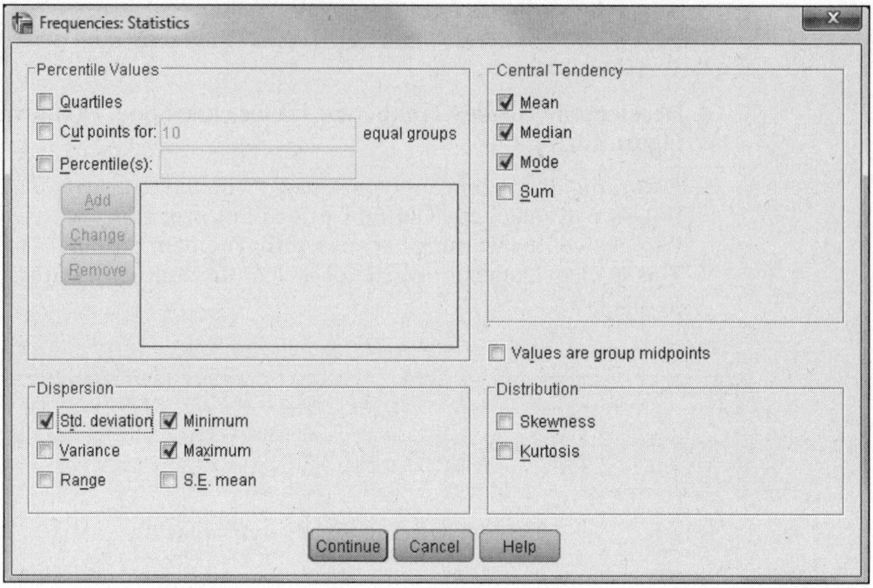

Figure 13-11:
The
Frequencies:
Statistics
dialog box.

7. **Click Continue.**

8. **Click the Charts button.**

   The Frequencies: Charts dialog box appears.

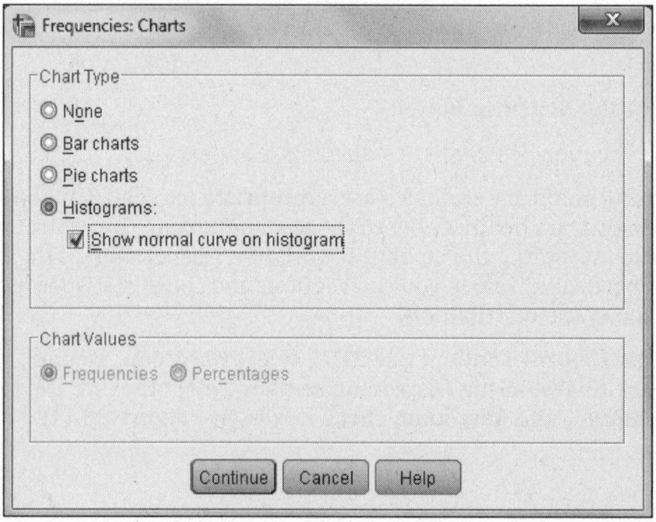

Figure 13-12:
The
Frequen-
cies: Charts
dialog box.

9. **Click the Histograms radio button and select the Show Normal Curve on Histogram check box, as shown in Figure 13-12.**

10. **Click Continue.**

11. **Click OK.**

SPSS runs the frequencies procedure and calculates the summary statistics and the histogram you requested.

The Statistics table (shown in Figure 13-13) shows that we have 3,338 valid cases and we don't have any missing data. The Statistics table contains the requested statistics. For example, for the variable Speakers, we can see that the minimum value is 0 and the maximum value is 451. This seems like a very large range of values, so it would be useful to double-check the data to make sure there are no errors. Likewise, in an ideal world, we would like the mean, median, and mode to be similar, because they're all measures of central tendency. In this case, note that for the variable Speakers, the mean (51.3), median (36), and mode (4) are very different from each other, which is an indication that this variable is probably not normally distributed (you see why this is important in later chapters).

Statistics

		Stereos	TVs	Speakers	Delivery_Prob lems	Years_as_cu stomer	Estimated_R evenue
N	Valid	3338	3338	3338	3338	3338	3338
	Missing	0	0	0	0	0	0
	Mean	13.71	.83	51.30	.13	6.38	5034510.94
	Median	14.00	.00	36.00	.00	6.00	5029070.00
	Mode	0	0	4	0	7	5029070
	Std. Deviation	9.417	2.228	54.104	.434	2.565	2828800.406
	Minimum	0	0	0	0	2	11028
	Maximum	30	10	451	4	11	9983290

Figure 13-13: The Statistics table.

You can visually check the distribution of these variables with a histogram (see Figure 13-14). A histogram has bars, but, unlike the bar chart, they're plotted along an equal interval scale. The height of each bar is the count of values falling within the interval. Notice that the lower range of values is truncated at 0 and the number of speakers is greatest down toward the lower end of the distribution, although there are some extreme values. The distribution is not normal.

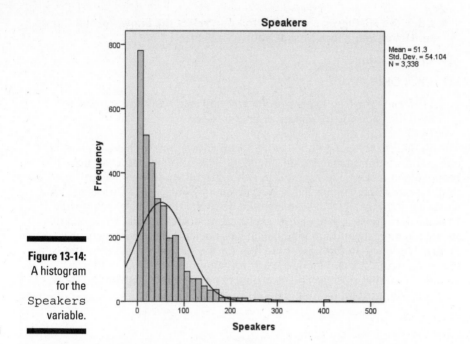

Figure 13-14:
A histogram
for the
Speakers
variable.

Summarizing Continuous Variables with the Descriptives Procedure

The descriptive procedure is an alternative to the frequencies procedure (see the preceding section) when the objective is to summarize continuous variables. The descriptives procedure provides a succinct summary of various statistics and the number of cases with valid values for each variable included in the table. To use the descriptives procedure, follow these steps:

1. **From the main menu, choose File ⇨ Open ⇨ Data and load the merchandise.sav data file.**

   The file is not in the SPSS installation directory. You have to download it from this book's companion website.

2. **Choose Analyze ⇨ Descriptive Statistics ⇨ Descriptives.**

   The Descriptives dialog box, shown in Figure 13-15, appears.

3. **Select the variables Stereos, TVs, Speakers, Delivery_ Problems, Years_as_customer, and Estimated_Revenue, and place them in the Variable(s) box, as shown in Figure 13-16.**

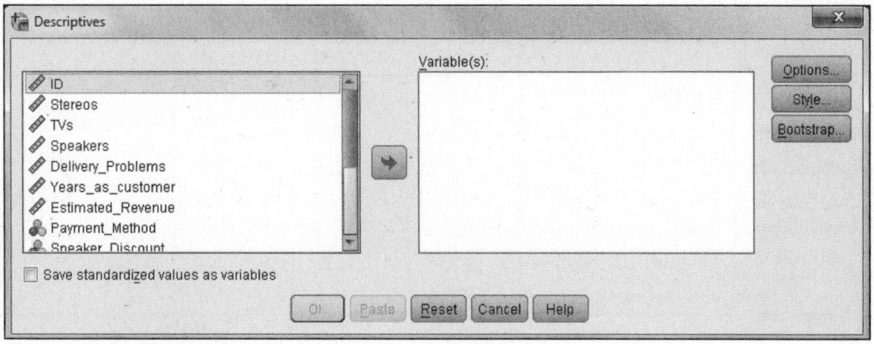

Figure 13-15:
The
Descriptives
dialog box.

4. **Click OK.**

   SPSS runs the descriptives procedure and calculates the summary
   statistics.

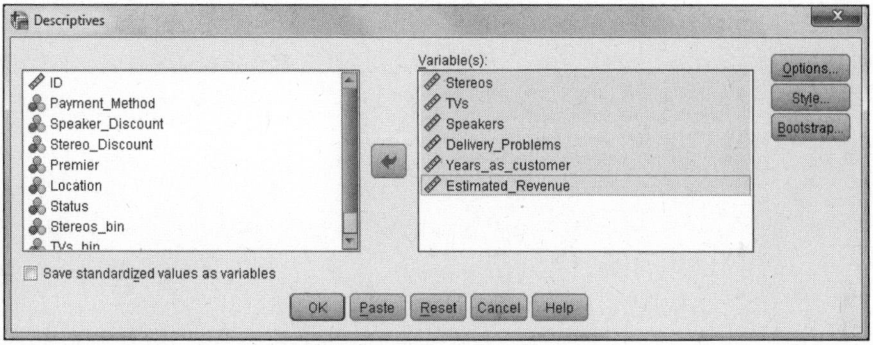

Figure 13-16:
Place the
variables
in the
Variable(s)
box.

The minimum and maximum values provide an efficient way to check for
values outside the expected range(see Figure 13-17). Likewise, it's always
important to investigate the mean and determine if the value makes sense.
Sometimes a mean may be lower or higher than expected, which can indicate a problem relating to how the data was coded or maybe even collected.
Finally, the last row in the table, Valid N (listwise), gives the number of cases
that have a valid value on all the variables appearing in the table. In this
example, we have no missing data, so this number isn't particularly useful
for this set of variables. However, it would be useful for a set of variables
that you intended to use for a *multivariate analysis* (an analysis looking at the
relationships between many variables).

Descriptive Statistics

	N	Minimum	Maximum	Mean	Std. Deviation
Stereos	3338	0	30	13.71	9.417
TVs	3338	0	10	.83	2.228
Speakers	3338	0	451	51.30	54.104
Delivery_Problems	3338	0	4	.13	.434
Years_as_customer	3338	2	11	6.38	2.565
Estimated_Revenue	3338	11028	9983290	5034510.94	2828800.406
Valid N (listwise)	3338				

Figure 13-17:
The Descriptive Statistics table.

Chapter 14

Showing Relationships between Categorical Dependent and Independent Variables

* *

In This Chapter

▶ Testing hypotheses

▶ Running inferential tests

▶ Running the crosstabs procedure

▶ Running the chi-square test

▶ Comparing column proportions

▶ Adding control variables

* *

Descriptive statistics (introduced in Chapter 13) describe the data in a sample through a number of summary procedures and statistics. For example, maybe you collected customer satisfaction data on a subset of your customers and you can determine that the average satisfaction for your customers is 3.5 on a 5-point scale. You may want to take this information a step further, though. For example, you may want to determine if there is a difference in satisfaction between customers who bought Product A (3.6) and customers who bought Product B (3.3). The numbers aren't exactly the same, but are they *really* different or are the differences due to random variation? This is the type of question that inferential statistics can answer.

Inferential statistics allow you to infer the results from the sample on which you have data and apply it to the population that the sample represents. Understanding how to make inferences from a sample to a population is the basis of inferential statistics. This allows you to reach conclusions about the population without the need to study every single individual.

In this chapter, we begin by discussing the idea of hypothesis testing. Next we run the crosstabs procedure to assess the relationship between two categorical variables. Finally, we use the chi-square test to see if there is a statistically significant relationship between two categorical variables.

Testing a Hypothesis to See If It's Right

Whenever you want to make an inference about a population from a sample, you must test a specific hypothesis. Typically, you state two hypotheses:

- **Null hypothesis:** The null hypothesis is conventionally the one in which no effect is present. For example, you may be looking for differences in mean income between males and females, but the (null) hypothesis you're testing is that there is no difference between the groups. Or the null hypothesis may be that there are no differences in satisfaction between customers who bought Product A (3.6) and customers who bought Product B (3.3); in other words, the differences are due to random variation.

- **Alternative hypothesis:** The alternative hypothesis is generally (although not exclusively) the one researchers are really interested in. For example, you may hypothesize that the mean incomes of males and females are different. Or for the customer satisfaction example, the alternative hypothesis may be that there is a difference in satisfaction between customers who bought Product A (3.6) and customers who bought Product B (3.3); in other words, the differences are real.

Now, in statistics, we never know anything for certain because we're dealing with samples, rather than populations. So, we always have to work with probabilities. The way hypotheses are assessed is by calculating the probability or the likelihood of finding our result. A probability value can range from 0 to 1 (corresponding to 0 percent to 100 percent, in terms of percentages); you can use these values to assess whether the likelihood that any differences you've found are the result of random chance.

So, how do hypotheses and probabilities interact? Suppose you want to know who will win the Super Bowl. You ask your fellow statisticians, and one of them says that he has built a predictive model and he knows Team A will win. Your next question should be, "How confident are you in your prediction?" Your friend says, "I'm 50 percent confident." Are you going to trust this prediction? Of course not, because there are only two outcomes and 50 percent is just random chance.

So, you ask another fellow statistician, and he tells you that he has built a predictive model. He knows that Team A will win, and he's 75 percent confident in his prediction. Are you going to trust his prediction? Well, now you start to think about it a little. You have a 75 percent chance of being right, and a 25 percent chance of being wrong. Let's say you decide a 25 percent chance of being wrong is too high.

So, you ask another fellow statistician and she tells you that she has built a predictive model and she knows Team A will win, and she's 90 percent confident in her prediction. Are you going to trust her prediction? Now you have a 90 percent chance of being right, and only a 10 percent chance of being wrong.

This is the way statistics work. You have two hypotheses — the null hypothesis and the alternative hypothesis — and you want to be sure of your conclusions. So, having formally stated the hypotheses, you must then select a criterion for acceptance or rejection of the null hypothesis. With probability tests, such as the chi-square test or the t-test, you're testing the likelihood that a statistic of the magnitude obtained (or greater) would've occurred by chance, assuming that the null hypothesis is true. You always assess the null hypothesis, which is the hypothesis that states there is no difference or no relationship. In other words, you only wish to reject the null hypothesis when you can say that the result would've been extremely unlikely under the conditions set by the null hypothesis. In this case, the alternative hypothesis should be accepted.

But what criterion (or alpha level, as it is often known) should you use? Traditionally, a 5 percent level is chosen, indicating that a statistic of the size obtained would only be likely to occur on 5 percent of occasions (or once in 20) should the null hypothesis be true. This also means that, by choosing a 5 percent criterion, you're accepting that you'll make a mistake in rejecting the null hypothesis 5 percent of the time, which most of the time you can live with.

Conducting Inferential Tests

As we mention in Chapter 13, obtaining descriptive statistics can be very important for various reasons. As mentioned in the previous chapter, one reason why descriptive statistics are important is because they form the basis for more complex analyses. In reality, when you run a chi-square test (see "Running the chi-square test," later in this chapter), a t-test (see Chapter 15), or a regression (see Chapter 16), all you're really doing is taking descriptive statistics and plugging them into a fancy formula, which then provides you with useful information.

Statistics are available for variables at all levels of measurement for more advanced analysis. In practice, the choice of method depends on the questions you're interested in asking of the data and the level of measurement of the variables involved. Table 14-1 suggests which statistical techniques are most appropriate, based on the measurement level of the dependent (effect) and independent (cause) variables.

Table 14-1	Level of Measurement and Statistical Tests	
Dependent Variables	*Independent Variables*	
	Categorical	*Continuous*
Categorical	Crosstabs	Logistic regression
Continuous	T-test, analysis of variance (ANOVA)	Correlation, linear Regression

In this chapter, we discuss crosstabs, which is when both the independent and dependent variables are categorical. In Chapter 15, we discuss t-tests, which is when the independent variable is categorical and the dependent variable is continuous. In Chapter 16, we discuss correlation and regression, which is when both the independent and dependent variables are continuous. We don't cover logistic regression, which is when the independent variable is continuous and the dependent variable is categorical, because that technique is beyond the scope of this book.

Running the Crosstabs Procedure

One of the most common ways to analyze data is to use crosstabulations. A crosstabulation is used when you want to study the relationship between two or more categorical variables. For example, you may want to look at the relationship between gender and handedness (whether you're right or left handed). This way, you can determine if one gender is more likely to be right or left handed, or if handedness is equally distributed between the genders. In this section, we provide examples and advice on how best to construct and interpret crosstabulations.

Here's how to perform a crosstabulation using the data file
`merchandise.sav`, which is not installed with SPSS Statistics:

1. **From the main menu, choose File ⇨ Open ⇨ Data and load the
   `merchandise.sav` data file, which is not in the IBM SPSS Statistics
   directory.**

   This file contains the customer's purchase history and has 16 variables
   and 3,338 cases.

2. **Choose Analyze ⇨ Descriptive Statistics ⇨ Crosstabs.**

   The Crosstabs dialog box (shown in Figure 14-1) appears.

   In this example, we want to study whether the number (`Small`, `Medium`,
   or `Large`) of TVs (`TVs_bin`) purchased this year is related to customer
   status (`Current` or `Churned`).

   You can place the variables in either the Rows or Columns boxes, but
   in many areas of research, it's customary to place the independent variable in the column of the table.

3. **Select the variable `Status`, and place it in the Row(s) box.**

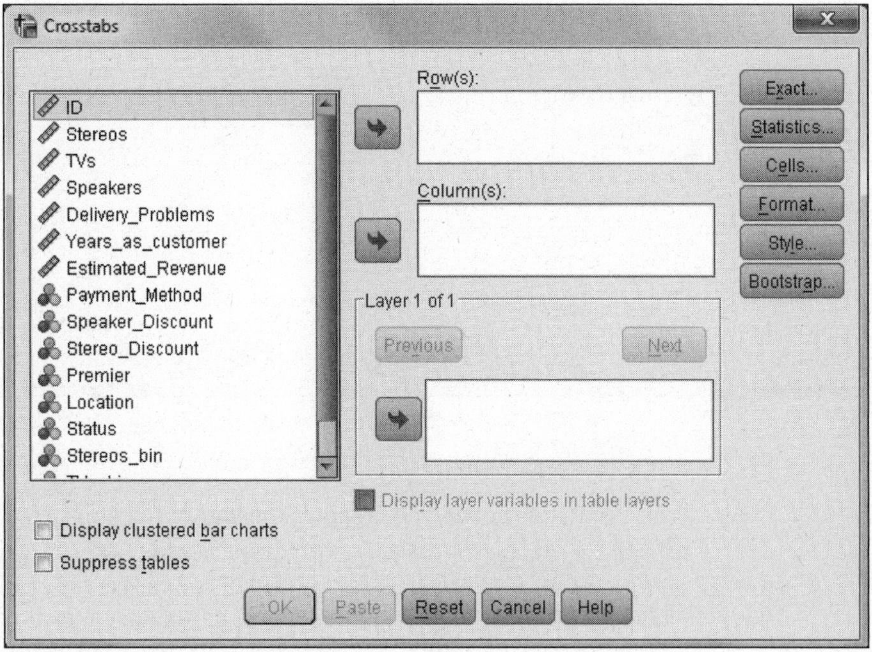

Figure 14-1:
The
Crosstabs
dialog box.

4. **Select the variable** `TVs_bin`, **and place it in the Column(s) box, as shown in Figure 14-2.**

As previously mentioned, crosstabulations are used when looking at the relationship between categorical variables, which is what we have in this situation.

In this example, only one variable was added to the Rows and Columns boxes, but you can add multiple variables to these boxes, which will create separate tables for all combinations of variables.

At this point, you can run the crosstabulation procedure, but you normally want to request additional statistics, typically percentages.

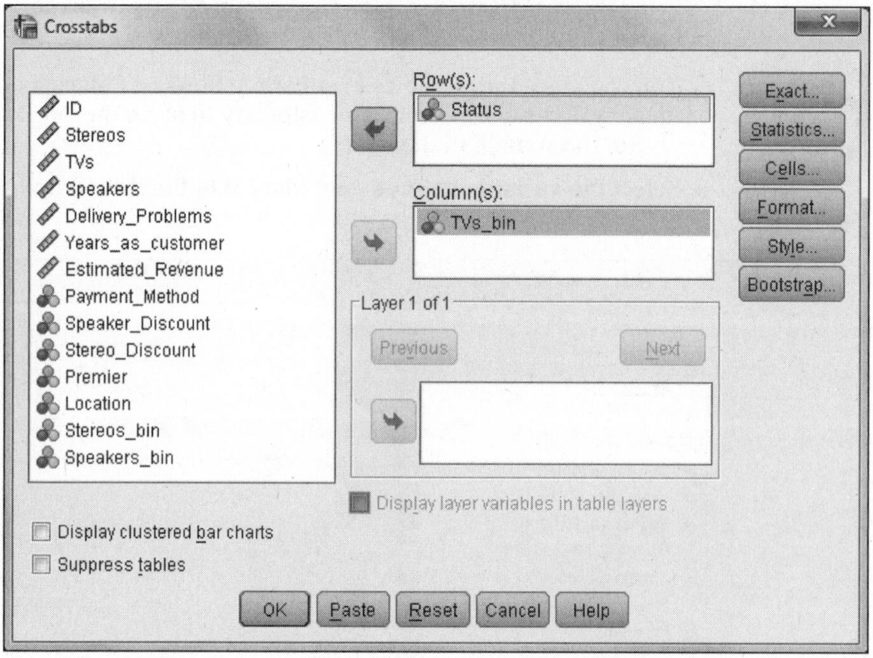

Figure 14-2:
The com-
pleted
Crosstabs
dialog box.

5. **Click the Cells button.**

The Cell Display dialog box (shown in Figure 14-3) appears.

By default, the cell count is displayed (that is, the Observed check box in the Counts area is selected). Typically, you would select row and/or column percents (by clicking the Row and Column check boxes in the Percentages area). Other less-used statistics, including residuals, are available in this dialog box as well.

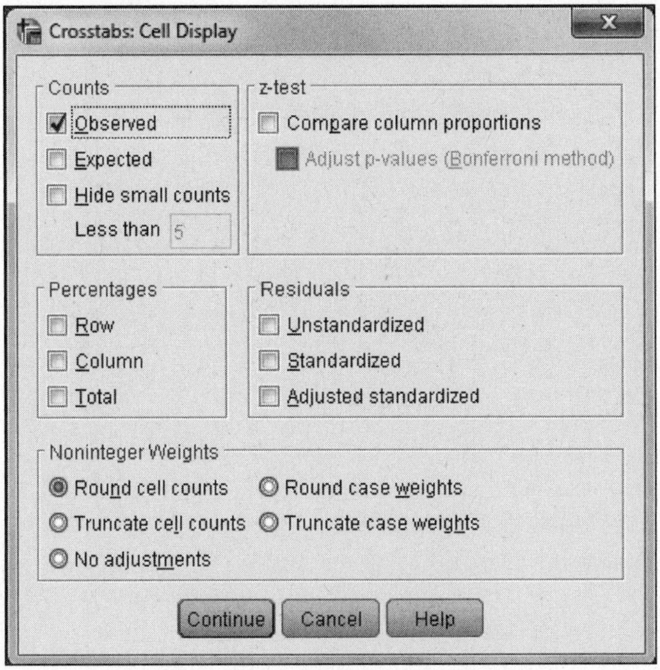

Figure 14-3:
The Cell
Display dia-
log box.

Crosstabs: Cell Display

Counts
- ☑ Observed
- ☐ Expected
- ☐ Hide small counts
  - Less than 5

z-test
- ☐ Compare column proportions
- ☐ Adjust p-values (Bonferroni method)

Percentages
- ☐ Row
- ☐ Column
- ☐ Total

Residuals
- ☐ Unstandardized
- ☐ Standardized
- ☐ Adjusted standardized

Noninteger Weights
- ⦿ Round cell counts
- ○ Truncate cell counts
- ○ No adjustments
- ○ Round case weights
- ○ Truncate case weights

Continue   Cancel   Help

6. **Click the Column check box in the Percentages area, as shown in Figure 14-4.**

7. **Click Continue.**

8. **Click OK.**

The Case Processing Summary table (shown in Figure 14-5) displays the number of valid and missing cases for the variables requested in the crosstabulation. Only the valid cases are displayed in the crosstabulation table.

Be sure to review this table to check the number of missing cases. If there are substantial amounts of missing data, you may want to question why this is the case and how your analysis will be affected. In this example, we don't have any missing data.

The crosstabulation table (shown in Figure 14-6) shows the relationship between the variables. Each cell of the table represents a unique combination of the variables values. For example, the first cell in the crosstabulation table shows the number of customers (1,093) who purchased a small number of TVs and also churned.

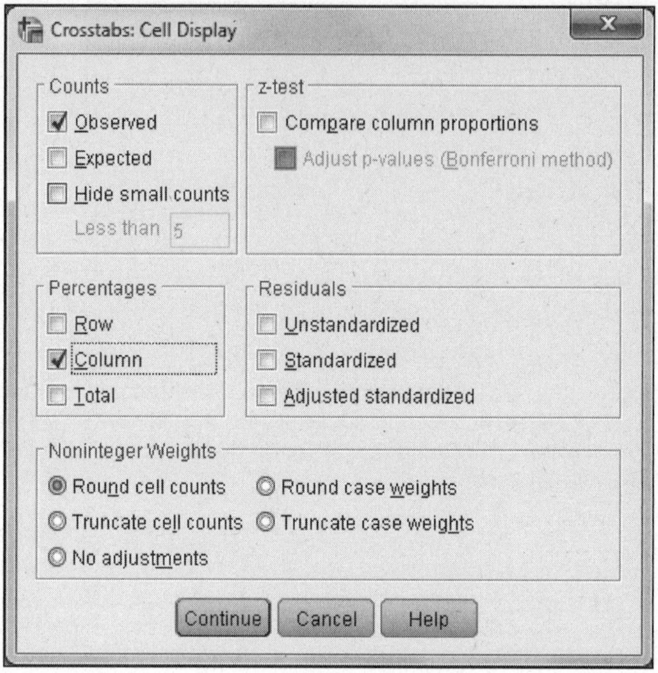

Figure 14-4:
The completed Cell Display dialog box.

Crosstabs: Cell Display

Counts
- ☑ Observed
- ☐ Expected
- ☐ Hide small counts
  - Less than 5

z-test
- ☐ Compare column proportions
- ☐ Adjust p-values (Bonferroni method)

Percentages
- ☐ Row
- ☑ Column
- ☐ Total

Residuals
- ☐ Unstandardized
- ☐ Standardized
- ☐ Adjusted standardized

Noninteger Weights
- ◉ Round cell counts
- ◉ Round case weights
- ○ Truncate cell counts
- ○ Truncate case weights
- ○ No adjustments

Continue | Cancel | Help

Figure 14-5:
The Case Processing Summary table.

Case Processing Summary

	Cases					
	Valid		Missing		Total	
	N	Percent	N	Percent	N	Percent
Status * Binned input variable TVs	3338	100.0%	0	0.0%	3338	100.0%

Figure 14-6:
The crosstabulation table.

Status * TVs_bin Binned input variable TVs Crosstabulation

			TVs_bin Binned input variable TVs			Total
			Small	Medium	Large	
Status	Churned	Count	1093	98	258	1449
		% within TVs_bin Binned input variable TVs	37.8%	96.1%	74.4%	43.4%
	Current	Count	1796	4	89	1889
		% within TVs_bin Binned input variable TVs	62.2%	3.9%	25.6%	56.6%
Total		Count	2889	102	347	3338
		% within TVs_bin Binned input variable TVs	100.0%	100.0%	100.0%	100.0%

TIP

Although looking at counts is useful, it's usually much easier to detect patterns by examining percentages. This is why you clicked the Column check box in the Cell Display dialog.

Looking at the first column in the crosstabulation table, you can see that 37.8% of customers who purchased a small number of TVs churned, while 62.2% of customers who purchased a small number of TVs stayed as customers. So it seems as though purchasing a few TVs is associated with staying as a customer. However, it seems that the opposite is true when more TVs are purchased — in other words, purchasing more TVs seems to be associated with losing customers, because we lost 96.1 percent and 74.4 percent of customers, respectively, who purchased a medium and large number of TVs.

These differences in percentages would certainly lead you to conclude that the number of TVs purchased is related to customer status. But how do you know that these differences in percentages are not due to chance? To answer this question we need to perform the chi-square test.

Running the chi-square test

The most common test used in a crosstabulation is the Pearson chi-square test, which tests the null hypothesis that the row and column variables are not related to each other — that is, that the variables are independent. In our situation, the chi-square test determines whether there is a relationship between number of TVs purchased and customer status.

To run the chi-square test, follow these steps:

1. **Choose Analyze ➪ Descriptive Statistics ➪ Crosstabs.**

   Remember that you should have the variable Status in the Row(s) box and the variable TVs_bin in the Column(s) box.

2. **Click the Statistics button.**

   The Statistics dialog box (shown in Figure 14-7) appears.

   A number of association measures are available in the Statistics dialog box. These measures of association characterize the relationship between the variables in the table. The measures are grouped in the dialog box by the measurement level of the variables in the table. They show the strength of relationship between the variables, whereas the chi-square test determines whether there is a statistically significant relationship.

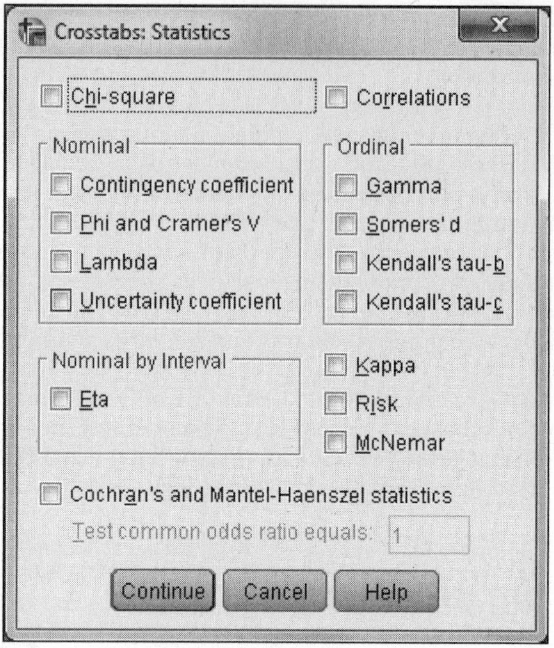

Figure 14-7:
The
Statistics
dialog box.

Crosstabs: Statistics

☐ Chi-square   ☐ Correlations

Nominal
☐ Contingency coefficient
☐ Phi and Cramer's V
☐ Lambda
☐ Uncertainty coefficient

Ordinal
☐ Gamma
☐ Somers' d
☐ Kendall's tau-b
☐ Kendall's tau-c

Nominal by Interval
☐ Eta

☐ Kappa
☐ Risk
☐ McNemar

☐ Cochran's and Mantel-Haenszel statistics
Test common odds ratio equals: 1

Continue   Cancel   Help

3. **Click the Chi-square check box, as shown in Figure 14-8.**

4. **Click Continue.**

5. **Click OK.**

   The same case processing summary and crosstabulation tables appear
   as shown earlier, but the Chi-Square Tests table (shown in Figure 14-9)
   also appears.

Three chi-square values are listed, the first two of which are used to test for
a relationship. Concentrate on the Pearson Chi-Square statistic, which is ade-
quate for almost all purposes. The Pearson Chi-Square statistic is calculated
by testing the difference between the *observed counts* (the number of cases
we actually observed in each crosstabulation cell) and the *expected counts*
(the number of cases we should've observed in each crosstabulation cell if
there were no relationship between the variables). So, the chi-square statistic
is an indication of misfit between observed minus expected counts.

You can request Expected counts in the Cell Display dialog box (refer to
Figure 14-3).

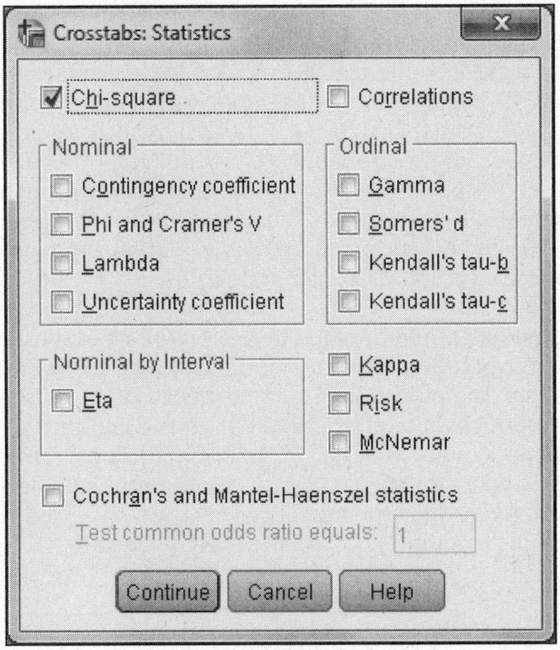

Figure 14-8:
Completed
Statistics
dialog box.

Chi-Square Tests

	Value	df	Asymptotic Significance (2-sided)
Pearson Chi-Square	286.989[a]	2	.000
Likelihood Ratio	308.196	2	.000
Linear-by-Linear Association	225.583	1	.000
N of Valid Cases	3338		

a. 0 cells (0.0%) have expected count less than 5. The minimum expected count is 44.28.

Figure 14-9:
The Chi-Square Tests table.

The actual chi-square value (here, 286.989) is used in conjunction with the number of degrees of freedom (df), which is related to the number of cells in the table, to calculate the significance for the chi-square statistic, labeled "Asymptotic Significance (2-sided)."

The significance value provides the probability of the null hypothesis being true, so the lower the number, the less likely that the variables are unrelated. Analysts often use a cutoff value of 0.05 or lower, to determine whether the results are statistically significant. For example, with a cutoff value of 0.05, if the significance value is smaller than 0.05, the null hypothesis is rejected. In this case, you can see that the probability of the null hypothesis being true is very small — in fact, it's less than 0.05, so you can reject the null hypothesis and you have no choice but to say you found support for the research hypothesis. So, you can conclude that there is a relationship between the number of TVs purchased and customer status.

Every statistical test has assumptions. The better you meet these assumptions, the more you can trust the results of the test. The chi-square test just assumes that you have a large enough sample. Because of this, there is a footnote to the Chi-Square Tests table (refer to Figure 14-9) noting the number of cells with expected counts less than 5. If more than 20 percent of the cells have this condition, you should consider increasing your sample size if you can, or else you should reduce the number of cells in your crosstabulation table. You can do this by either combining or removing categories.

Comparing column proportions

After determining that there is a relationship between two variables, the next step is to determine the exact nature of the relationship — that is, which groups actually differ from each other. You can see which groups differ from each other by comparing column proportions. To do this, follow these steps:

1. **Choose Analyze ⇨ Descriptive Statistics ⇨ Crosstabs.**

   Remember that you should have the variable Status in the Row(s) box and the variable TVs_bin in the Column(s) box.

2. **Click the Cells button.**

3. **Click the Compare Column Proportions check box.**

4. **Click the Adjust P-Values (Bonferroni Method) check box.**

   Figure 14-10 shows the completed Cell Display dialog box.

5. **Click Continue.**

6. **Click OK.**

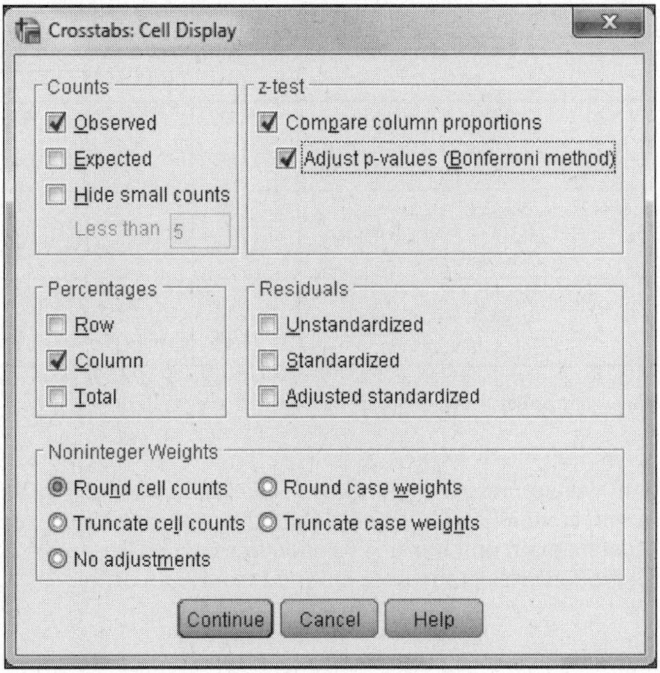

Figure 14-10:
The completed Cell
Display dialog box.

The same Case Processing Summary and Chi-Square Tests tables appear as shown earlier, but a modified version of the crosstabulation table now appears (see Figure 14-11).

The crosstabulation table now includes the column proportions test notations. Subscript letters are assigned to the categories of the column variable. For each pair of columns, the column proportions are compared using a z-test. If a pair of values is significantly different, different subscript letters are displayed in each cell.

The table in Figure 14-11 shows that the proportion of customers who purchased a small number of TVs and churned (37.8 percent) is smaller and significantly different, according to the z-test, than the proportion of customers who purchased a medium number of TVs and churned (96.1 percent), due to having different subscript letters. You can also see that the proportion of customers who purchased a small number of TVs and churned (37.8 percent) is significantly smaller than the proportion of customers who purchased a large number of TVs and churned (74.4 percent). Finally, we can see that the

Status * TVs_bin Binned input variable TVs Crosstabulation

			TVs_bin Binned input variable TVs			Total
			Small	Medium	Large	
Status	Churned	Count	1093a	98b	258c	1449
		% within TVs_bin Binned input variable TVs	37.8%	96.1%	74.4%	43.4%
	Current	Count	1796a	4b	89c	1889
		% within TVs_bin Binned input variable TVs	62.2%	3.9%	25.6%	56.6%
Total		Count	2889	102	347	3338
		% within TVs_bin Binned input variable TVs	100.0%	100.0%	100.0%	100.0%

Each subscript letter denotes a subset of TVs_bin Binned input variable TVs categories whose column proportions do not differ significantly from each other at the .05 level.

Figure 14-11: The cross-tabulation table with compare column proportions test.

proportion of customers who purchased a large number of TVs and churned (74.4 percent) is significantly smaller than the proportion of customers who purchased a medium number of TVs and churned (96.1 percent). In other words, all groups are significantly different from each other.

Adding control variables

Tables can be made more complex by adding variables to the layer dimension. This way, you can create layered tables displaying relationship among three or more variables. A layered variable further subdivides categories of the row and column variables by the categories of the layer variable(s). The layer variables are often referred to as control variables because they show the relationship between the row and column variables when you "control" for the effects of the third variable. Layer variables are usually added to a table after the two-variable crosstabulation table has been examined.

To add a layer variable, follow these steps:

1. **Choose Analyze ⇨ Descriptive Statistics ⇨ Crosstabs.**

   Remember that you should have the variable Status in the Row(s) box and the variable TVs_bin in the Column(s) box.

2. **Select the variable Location, and place it in the Layer box, as shown in Figure 14-12.**

3. **Click OK.**

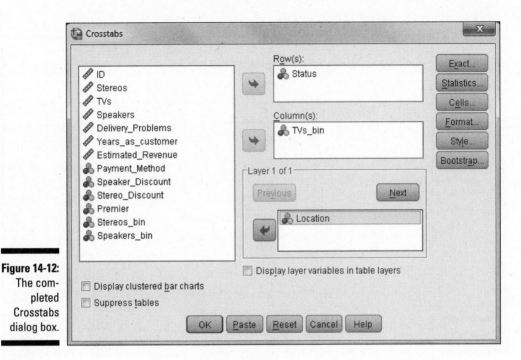

Figure 14-12:
The completed Crosstabs dialog box.

You can add more than one control variable to a table by using the Next button in the layer box, but you'll need to have a large sample size or you may quickly create tables with only a few cases, or even no cases, in several cells.

As shown in Figure 14-13, adding a layer variable creates subtables for each category of the layer variable (in this case, we have a table for international and national customers). In this example, there doesn't appear to be a difference in the relationship between the number of TVs purchased and customer status.

Figure 14-14 further confirms that the relationship between the number of TVs purchased and customer status is not affected by location. In other situations, however, a layer variable often helps qualify a relationship, so we know that a relationship only holds in certain conditions and not others.

Status * TVs_bin Binned input variable TVs * Location Crosstabulation

Location				Small	Medium	Large	Total
				TVs_bin Binned input variable TVs			Total
International	Status	Churned	Count	463a	27b	109b	599
			% within TVs_bin Binned input variable TVs	39.1%	87.1%	73.2%	43.9%
		Current	Count	721a	4b	40b	765
			% within TVs_bin Binned input variable TVs	60.9%	12.9%	26.8%	56.1%
	Total		Count	1184	31	149	1364
			% within TVs_bin Binned input variable TVs	100.0%	100.0%	100.0%	100.0%
National	Status	Churned	Count	630a	71b	149c	850
			% within TVs_bin Binned input variable TVs	37.0%	100.0%	75.3%	43.1%
		Current	Count	1075a	0b	49c	1124
			% within TVs_bin Binned input variable TVs	63.0%	0.0%	24.7%	56.9%
	Total		Count	1705	71	198	1974
			% within TVs_bin Binned input variable TVs	100.0%	100.0%	100.0%	100.0%
Total	Status	Churned	Count	1093a	98b	258c	1449
			% within TVs_bin Binned input variable TVs	37.8%	96.1%	74.4%	43.4%
		Current	Count	1796a	4b	89c	1889
			% within TVs_bin Binned input variable TVs	62.2%	3.9%	25.6%	56.6%
	Total		Count	2889	102	347	3338
			% within TVs_bin Binned input variable TVs	100.0%	100.0%	100.0%	100.0%

Figure 14-13: The cross-tabulation table with a layer variable.

Each subscript letter denotes a subset of TVs_bin Binned input variable TVs categories whose column proportions do not differ significantly from each other at the .05 level.

Chi-Square Tests

Location		Value	df	Asymptotic Significance (2-sided)
International	Pearson Chi-Square	86.313[b]	2	.000
	Likelihood Ratio	88.760	2	.000
	Linear-by-Linear Association	74.856	1	.000
	N of Valid Cases	1364		
National	Pearson Chi-Square	203.538[c]	2	.000
	Likelihood Ratio	230.678	2	.000
	Linear-by-Linear Association	152.732	1	.000
	N of Valid Cases	1974		
Total	Pearson Chi-Square	286.989[a]	2	.000
	Likelihood Ratio	308.196	2	.000
	Linear-by-Linear Association	225.583	1	.000
	N of Valid Cases	3338		

Figure 14-14: The Chi-Square Tests table with a layer variable.

a. 0 cells (0.0%) have expected count less than 5. The minimum expected count is 44.28.

b. 0 cells (0.0%) have expected count less than 5. The minimum expected count is 13.61.

c. 0 cells (0.0%) have expected count less than 5. The minimum expected count is 30.57.

Chapter 15

Showing Relationships between Continuous Dependent and Categorical Independent Variables

....

....

*I*n this chapter, we explain how to compare different groups on a continuous outcome variable. For example, we can compare residents of two cities on how much money they spend on rent to determine if there are significant differences.

But before we dive into this topic, we need to tackle another issue: In Chapter 14, you get your feet wet on the idea of hypothesis testing. In this chapter, we extend that discussion.

Hypothesis Testing Revisited

In Chapter 14, we introduce the idea of inferential statistics and mention how they allow us to infer the results from our sample to the population. This concept is important because we want to do research that applies to a larger audience than just the specific group of people we tested.

In Chapter 14, we also mention how hypothesis testing allows researchers to develop hypotheses, which are then assessed to determine the probability or likelihood of the findings. Two hypotheses are typically created:

- ✔ **Null hypothesis:** The null hypothesis states that no effect is present.

- ✔ **Alternative hypothesis:** The alternative hypothesis states that an effect is present.

For example, you may be interested in assessing if there are differences in mean income between males and females; the null hypothesis states that there is no difference between the groups, and the alternative hypothesis states that there is a difference between the groups.

The null hypothesis is then assessed by calculating the probability that the null hypothesis is true. At this point, we investigate the probability value, and if it's less than 0.05, we say that we've found support for the alternative hypothesis because the probability that the null hypothesis is true is low (less than 5%). If it's greater than 0.05, we say that we've found support for the null hypothesis because there is a decent chance that it's true (greater than 5%).

However, too often people immediately jump to the conclusion that the finding "is statistically significant" or "is not statistically significant." Although that's literally true, because we use those words to describe probability values below 0.05 and above 0.05, it doesn't imply that there are only two conclusions to draw about our finding. Table 15-1 is a more realistic scenario.

Table 15-1	Types of Statistical Outcome	
In the Real World	*Statistical Test Outcome*	
	Not Significant	**Significant**
No Difference (Null Is True)	Correct decision	False positive, used wrong test
True Difference (Alternative Is True)	False negative, used wrong test	Correct decision

Notice that in this table, several outcomes are possible. Let's take a look at the first row. It could be that, in the real world, there is no relationship between the variables, and this is what our test found. In this scenario, we would be making a correct decision. However, what if in the real world there were no relationship between the variables, and our test found that there was a significant relationship? In this case, we would be making an error; this type of error is called a *false positive* because the researcher falsely concludes a positive result (thinks it does occur). This type of error is explicitly taken into account when performing statistical tests. When testing for statistical significance using a 0.05 criterion, you acknowledge that if there is no effect in the population, the sample statistic will exceed the criterion on average 5 times out of 100 (or 0.05). So, this type of error could occur strictly by chance, but it could also occur if the researcher uses the wrong test. (An inappropriate test is used when you don't meet the assumptions of a test, which is why knowing and testing assumptions is important.)

How could SPSS let you do the "wrong test"? The calculations will always be correct, of course, but you have to know which menu to work in. For instance, the T-test is a parametric test, and it assumes that your data is shaped like a bell curve. In fact (and this is even more technical), it assumes that the "errors" are shaped like a bell curve. And there is a whole separate menu with tests that you can use when this isn't true. Whew! Don't worry about it too much because when you're expected to know about lots of alternative tests, you'll know. You may be surprised how many SPSS users get confused about these issues. What happens is this: A parametric test might yield a probability of 0.047 and a nonparametric test might yield a probability of 0.053. You can see the problem now. If you declare "significant," an expert might say "not significant."

For now, the main message is this — assumptions are not just a bunch of arbitrary rules. They sometimes affect which conclusion you draw.

Now let's take a look at the second row of Table 15-1. It could be that in the real world there is a relationship between the variables, and this is what our test found. In this scenario, we would be making a correct decision. However, what if in the real world there were a relationship between the variables, and our test found that there was no significant relationship? In this case, we would be making an error; this type of error is called a *false negative* because the researcher falsely concludes a negative result (thinks it does not occur). This type of error typically happens when we use small samples, so our test is not powerful enough to detect true differences. The error happens because, when sample sizes are small, precision tends to be poor. The error could also occur if the researcher uses the wrong test.

Using the Compare Means Dialog Box

Often, you encounter situations where you have a continuous dependent variable and a categorical independent variable. For example, you may want to determine if there are differences between the genders on SAT scores.

Or you may want to see if there was a change in behavior by assessing some-one before and after an intervention. In both of these examples, you're com-paring groups on some continuous outcome measure, and the statistic you're using to make comparisons among groups is the mean. For these kinds of analyses, you want to use the procedures that are within the Compare Means dialog box (shown in Figure 15-1). The Compare Means dialog box (which is accessed by choosing Analyze ⇨ Compare Means) contains six statistical techniques that allow users to compare sample means:

> M Means...
>
> t One-Sample T Test...
>
> 🔡 Independent-Samples T Test...
>
> ➕ Summary Independent-Samples T Test
>
> 🔡 Paired-Samples T Test...
>
> 🔡 One-Way ANOVA...

- ✔ **Means:** Calculates subgroup means and related statistics for dependent variables within categories of one or more independent variables.

- ✔ **One-Sample T Test:** Tests whether the mean of a single variable dif-fers from a specified value (for example, a group using a new learning method compared to the school average).

- ✔ **Independent-Samples T Test:** Tests whether the means for two groups differ on a continuous dependent variable (for example, females versus males on income).

- ✔ **Summary Independent-Samples T Test:** Uses summary statistics to test whether the means for two groups differ on a continuous dependent variable (for example, females versus males on income).

- ✔ **Paired-Samples T Test:** Tests whether means differ from each other under two conditions (for example, before versus after an intervention, or standing versus sitting).

- ✔ **One-Way ANOVA:** Tests whether the means for two or more groups differ on a continuous dependent variable (for example, drug1 versus drug2 versus drug3 on depression).

Running the Independent-Samples T-Test Procedure

In this chapter, we focus on the independent-samples t test, which allows you to compare two different groups on a continuous dependent variable. For example, we may be comparing two different learning methods to determine their effect on math skills.

Here's how to perform an independent-samples t test:

1. **From the main menu, choose File ⇨ Open ⇨ Data and load the `employee_data.sav` file.**

   The file is not in the SPSS installation directory. You have to download it from this book's companion website.

   This file contains the employee information from a bank in the 1960s and has 10 variables and 474 cases.

2. **Choose Analyze ⇨ Compare Means ⇨ Independent-Samples T Test.**

   The Independent-Samples T Test dialog box, shown in Figure 15-2, appears.

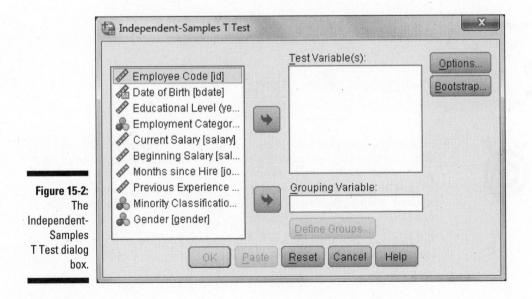

Figure 15-2: The Independent-Samples T Test dialog box.

3. **Select the variables** `current salary, beginning salary, months on the job, years of education,` **and** `previous job experience,` **and place them in the Test Variable(s) box.**

You can also place the categorical independent variable in the Grouping Variable box. Notice that you can have only one independent variable and that the program requires you to indicate which groups are to be compared.

4. **Select the variable** `gender` **and place it in the Grouping Variable box.**

5. **Click the Define Groups button.**

The Define Groups dialog box appears.

6. **In the Group 1 box, type 0; in the Group 2 box, type 1 (see Figure 15-3). Then click Continue.**

You're returned to the Independent-Samples T Test dialog box (shown in Figure 15-4).

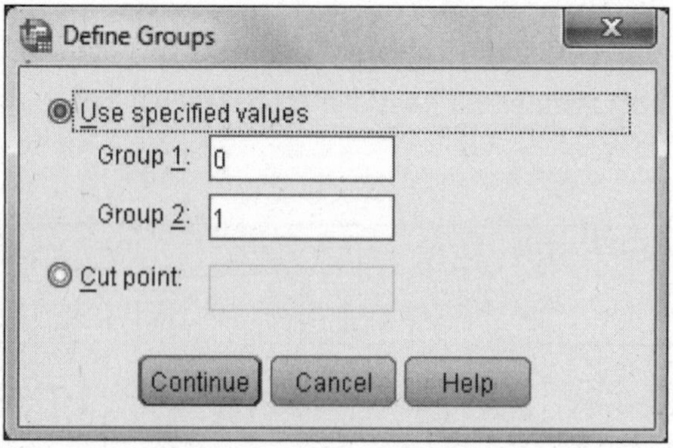

Figure 15-3:
The Define
Groups
dialog box.

If the independent variable is continuous, you can specify a cut point value to define the two groups. Those cases less than or equal to the cut point go into the first group, and those cases greater than the cut point fall into the second group. Also, if the independent variable is categorical but has more than two categories, it can be still used by specifying only two categories in an analysis.

You can also click the Options button and decide how to treat missing values and confidence intervals.

7. **Click OK.**

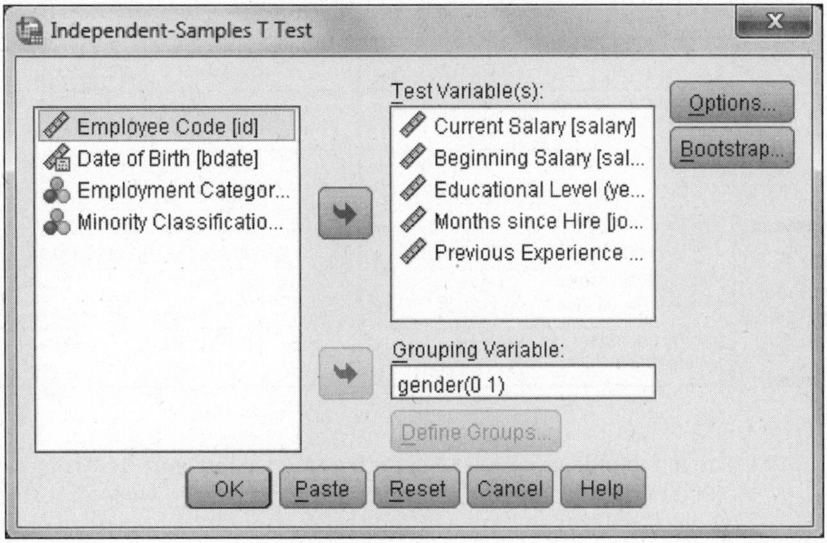

Figure 15-4:
The
Independent-
Samples
T Test dialog
box, with the
Grouping
Variable box
completed.

Every statistical test has assumptions. The better you meet these assumptions, the more you can trust the results of the test. The independent-samples t test has four assumptions:

✔ You're comparing only two groups.

✔ The dependent variable is continuous.

✔ The dependent variable is normally distributed within each category of the independent variable (normality).

✔ You have similar variation within each category of the independent variable (homogeneity of variance).

The Group Statistics table (shown in Figure 15-5) provides sample sizes, means, standard deviations, and standard errors for the two groups. You can see that there are a few more males (258) than females (216) in this sample; however, we're certainly not dealing with small sizes here.

The independent-samples t test is robust to moderate in terms of violations of the assumption of normality when sample sizes are moderate to large (more than 50 cases per group).

We can also see that males have higher means than females on all the dependent variables. This is precisely what the independent-samples t test will assess — whether the differences between the means are significantly different from each other or if the differences we're seeing are just due to chance.

Group Statistics

	gender Gender	N	Mean	Std. Deviation	Std. Error Mean
educ Educational Level (years)	Male	258	14.43	2.979	.185
	Female	216	12.37	2.319	.158
salary Current Salary	Male	258	$41,441.78	$19,499.214	$1,213.968
	Female	216	$26,031.92	$7,558.021	$514.258
salbegin Beginning Salary	Male	258	$20,301.40	$9,111.781	$567.275
	Female	216	$13,091.97	$2,935.599	$199.742
jobtime Months since Hire	Male	258	81.72	10.351	.644
	Female	216	80.38	9.676	.658
prevexp Previous Experience (months)	Male	258	111.62	109.692	6.829
	Female	216	77.04	95.012	6.465

Figure 15-5:
The Group Statistics table.

The Independent Samples Test table (shown in Figure 15-6) displays the result of the independent-samples t test, but before we look at this test, we must determine if we met the assumption of homogeneity of variance. Violating the assumption of homogeneity of variance is more critical than violating the assumption of normality, because when the former occurs, the significance or probability value reported by SPSS are incorrect and the test statistics must be adjusted.

Levene's test for equality of variances assesses the assumption of homogeneity of variance. This test determines if the variation is similar or different between the groups. When Levene's test is not statistically significant, it means that we can continue with the regular independent-samples t test. In this situation, we can look at the independent-samples t test table row that specifies that equal variances are assumed. When Levene's test is statistically significant, it means that there are differences in variation between the groups, so we have to make an adjustment to our independent-samples t test. In this situation, we can look at the independent-samples t test table row that specifies that equal variances are not assumed.

In the left section of the Independent Samples Test table, Levene's test for equality of variances is displayed. The F column displays the actual test result, which is used to calculate the significance level (the Sig. column).

In our example, we did meet the assumption of homogeneity of variance for the variables Months since hire and Previous experience, because the value in the Sig. column was greater than 0.05, so we can look at the row that specifies that equal variances are assumed. However, we did not meet the assumption of homogeneity of variance for the variables Education level, Current salary, and Beginning salary, because the value in the Sig. column was less than 0.05, so we can look at the row that specifies that equal variances are not assumed.

Independent Samples Test

		Levene's Test for Equality of Variances		t-test for Equality of Means						95% Confidence Interval of the Difference	
		F	Sig.	t	df	Sig. (2-tailed)	Mean Difference	Std. Error Difference		Lower	Upper
educ Educational Level (years)	Equal variances assumed	17.884	.000	8.276	472	.000	2.060	.249		1.571	2.549
	Equal variances not assumed			8.458	469.595	.000	2.060	.244		1.581	2.538
salary Current Salary	Equal variances assumed	119.669	.000	10.945	472	.000	$15,409.862	$1,407.906		$12,643.322	$18,176.401
	Equal variances not assumed			11.688	344.262	.000	$15,409.862	$1,318.400		$12,816.728	$18,002.996
salbegin Beginning Salary	Equal variances assumed	105.969	.000	11.152	472	.000	$7,209.428	$646.447		$5,939.158	$8,479.698
	Equal variances not assumed			11.987	318.818	.000	$7,209.428	$601.413		$6,026.188	$8,392.667
jobtime Months since Hire	Equal variances assumed	2.168	.142	1.447	472	.148	1.341	.927		-.480	3.162
	Equal variances not assumed			1.456	466.269	.146	1.341	.921		-.469	3.152
prevexp Previous Experience (months)	Equal variances assumed	2.582	.109	3.631	472	.000	34.583	9.524		15.889	53.297
	Equal variances not assumed			3.678	471.444	.000	34.583	9.404		16.105	53.062

Figure 15-6:
Independent
Samples
Test Table.

Now that we've determined whether we met the assumption of homogeneity of variance, we're ready to see if the differences between the means are significantly different from each other or if the differences we're seeing are just due to chance. The t column displays the actual result of the t test and the df column tells SPSS Statistics how to determine the probability of the t statistic. The Sig. (2-tailed) column tells us the probability of the null hypothesis being correct. If the probability value is very low (less than 0.05), we can conclude that the means are significantly different from each other. In Figure 15-6, you can see that there are significant differences between the males and females on all the variables except Months since hire.

An additional piece of useful information is the 95% confidence interval for the population mean difference. Technically, this tells you that if you were to continually repeat this study, you would expect the true population difference to fall within the confidence intervals 95% of the time. From a more practical standpoint, the 95% confidence interval provides a measure of the precision with which the true population difference is estimated. In the example, the 95% confidence interval for the mean difference between groups on years of education is from 1.581 to 2.538 years. The 95% confidence interval indicates the likely range within which you expect the population mean difference to fall. Notice that the difference values do not include zero, because there is a difference between groups. If zero had been included within the range, this would indicate that there are no differences between the groups — that is, you're saying that the probability value is greater than 0.05. In essence, the 95% confidence interval is another way of testing the null hypothesis. So, if the value of zero does not fall within the 95% confidence, you're saying that the probability of the null hypothesis (that is, no difference or a difference of zero) is less than 0.05.

Running the Summary Independent-Samples T-Test Procedure

The summary independent-samples t test procedure was added to SPSS in version 23. This procedure, like the independent-samples t-test procedure, performs an independent-samples t test. Here's the difference between these two procedures: In the independent-samples t-test procedure, you're required to have all your data to run the analysis (SPSS is reading your data directly from the data editor), whereas in the summary independent-samples t-test procedure, you just need to specify the number of cases for each group, the means, and the standard deviations to run the analysis.

The summary independent-samples t-test procedure could be very useful when analyzing data that you've collected, but it would be too time consuming to enter all the data into the SPSS Data Editor window. With this technique, you can quickly obtain means and standard deviations using a calculator and then run a t test. Or you can use this procedure to replicate the results of a published study for which you don't have the data.

To perform the summary independent-samples t-test procedure, follow these steps:

1. **From the main menu, choose File ⇨ Open ⇨ Data and load the** `employee_data.sav` **file.**

   The file is not in the SPSS installation directory. You have to download it from this book's companion website.

2. **Choose Analyze ⇨ Compare Means ⇨ Summary Independent-Samples T Test.**

   The T Test Computed from Summary Data dialog box, shown in Figure 15-7, appears.

   In this example, we want to study whether there are differences between two samples. You need to specify the number of cases in each group, along with the respective means and standard deviations.

3. **In the Number of Cases boxes under both Sample 1 and Sample 2, type 50.**

4. **In the Mean box for Sample 1, type 4.2; in the Mean box for Sample 2, type 3.5.**

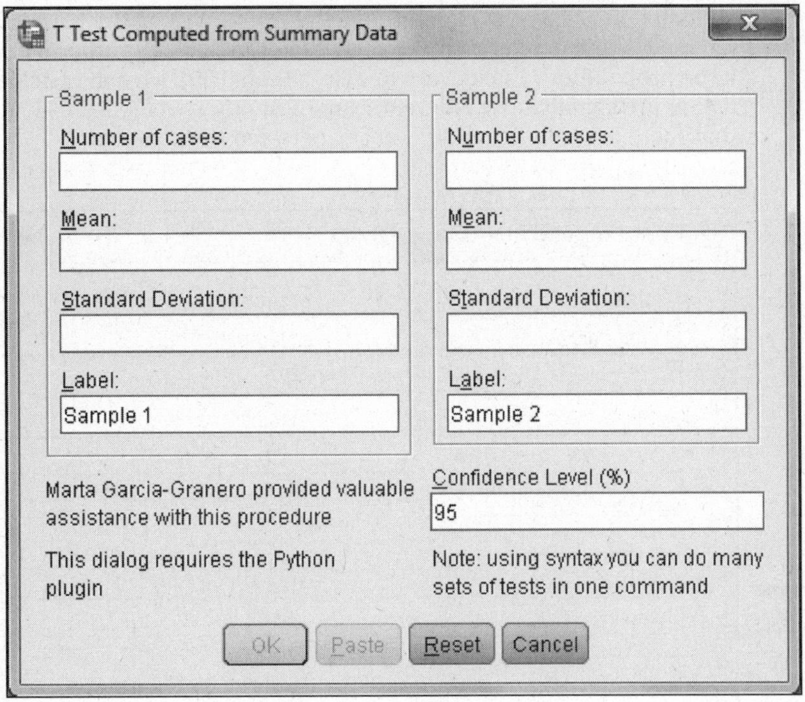

Figure 15-7:
The T Test
Computed
from
Summary
Data dialog
box.

5. In the Standard Deviation box for Sample 1, type 1.6; in the Standard Deviation box for Sample 2, type 1.5.

The completed dialog box is shown in Figure 15-8.

6. Click OK.

SPSS calculates the summary independent-samples t test.

The Summary Data table (shown in Figure 15-9) provides sample sizes, means, standard deviations, and standard errors for the two groups. This was the information we provided SPSS to perform this analysis.

The Independent Samples Test table (shown in Figure 15-10) displays the result of the independent-samples t test, along with the Hartley test of equal variance (which assesses the assumption of homogeneity of variance, like Levene's test). In our example, we did meet the assumption of homogeneity of variance because the Sig. value is greater than 0.05, so we can look at the row that specifies that equal variances are assumed. At this point, we're ready to see if the differences between the means are significantly different from each other or if the differences we're seeing are just due to chance. As before, the t column displays the actual result of the t test and the df column tells SPSS how to determine the probability of the t statistic. The Sig. (2-tailed) column tells us the probability of the null hypothesis being correct. If the probability value is very low (less than 0.05), we can conclude that the means are significantly different from each other. In Figure 15-10, we can see that there are significant differences between the samples.

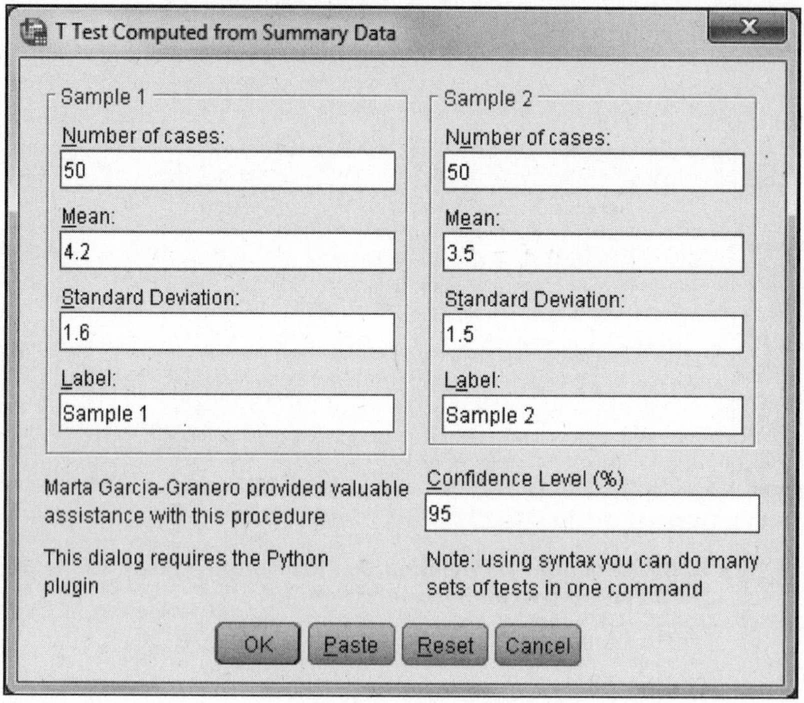

Figure 15-8:
The completed
T Test
Computed
from
Summary
Data dialog
box.

T Test Computed from Summary Data

Sample 1
Number of cases: 50
Mean: 4.2
Standard Deviation: 1.6
Label: Sample 1

Sample 2
Number of cases: 50
Mean: 3.5
Standard Deviation: 1.5
Label: Sample 2

Marta Garcia-Granero provided valuable assistance with this procedure

This dialog requires the Python plugin

Confidence Level (%)
95

Note: using syntax you can do many sets of tests in one command

OK   Paste   Reset   Cancel

Figure 15-9:
The
Summary
Data table.

Summary Data

	N	Mean	Std. Deviation	Std. Error Mean
Sample 1	50.000	4.200	1.600	.226
Sample 2	50.000	3.500	1.500	.212

Figure 15-10:
The
Independent
Samples
Test table.

Independent Samples Test

	Mean Difference	Std. Error Difference	t	df	Sig. (2-tailed)
Equal variances assumed	.700	.310	2.257	98.000	.026
Equal variances not assumed	.700	.310	2.257	97.595	.026

Hartley test for equal variance: F = 1.138, Sig. = 0.3249

The final table, 95.0% Confidence Intervals for Difference (shown in Figure 15-11), displays the 95% confidence intervals for the population mean difference. As we mention earlier, confidence intervals tell you that if you were to continually repeat this study, you would expect the true population difference to fall within the confidence intervals 95% of the time. Similar to the previous example, we have confidence intervals for when we meet or do not meet the assumption of homogeneity of variance. However, unlike in the previous example, here we have the option to use either the Asymptotic (an approximation) or Exact (the exact value) confidence intervals, whereas in the previous example we had to use the Asymptotic confidence intervals. The Exact 95% confidence interval for the mean difference between groups is from 0.84 to 1.316 (refer to Figure 15-11). Because the difference values don't include zero, there is a difference between groups.

95.0% Confidence Intervals for Difference

	Lower Limit	Upper Limit
Asymptotic (equal variance)	.092	1.308
Asymptotic (unequal variance)	.092	1.308
Exact (equal variance)	.084	1.316
Exact (unequal variance)	.084	1.316

Figure 15-11: The 95.0% Confidence Intervals for Difference table.

Chapter 16

Showing Relationships between Continuous Dependent and Independent Variables

. . . .

In This Chapter

▶ Viewing relationships

▶ Running the bivariate procedure

▶ Running the linear regression procedure

▶ Making predictions

. . . .

The two most commonly used statistical techniques to analyze relationships between continuous variables are the Pearson correlation and linear regression.

Many people use the term *correlation* to refer to the idea of a relationship between variables or a pattern. This view of the term *correlation* is correct, but correlation also refers to a specific statistical technique. Pearson correlations are used to study the relationship between two continuous variables. For example, you may want to look at the relationship between height and weight, and you may find that as height increases, so does weight. In other words, in this example, the variables are correlated with each other because changes in one variable impact the other.

Whereas correlation just tries to determine if two variables are related, linear regression takes this one step further and tries to predict the values of one variable based on another (so if you know someone's height, you can make an intelligent prediction for that person's weight). Of course, most of the time you wouldn't make a prediction based on just one independent variable (height); instead, you would typically use several variables that you deemed important (age, gender, BMI, and so on).

This chapter does not address how to create scatterplots, because we cover those in Chapter 12. However, you need to create scatterplots before using the correlation and linear regression procedures because these techniques are only appropriate when you have linear relationships.

Running the Bivariate Procedure

Correlations determine the similarity or difference in the way two continuous variables change in value from one case (row) to another through the data. As you can see in Figure 16-1, a scatterplot visually shows the relationship between two continuous variables by displaying individual observations. (This example uses the `employee_data.sav` data file.)

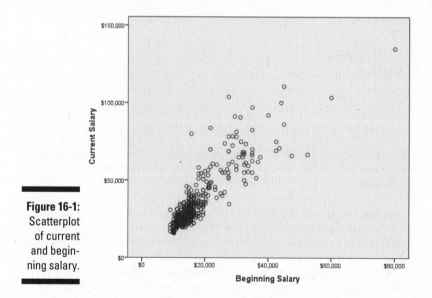

Figure 16-1:
Scatterplot of current and beginning salary.

Notice that, for the most part, low beginning salaries are associated with low current salaries, and that high beginning salaries are associated with high current salaries — this is called a *positive relationship*. Positive relationships show that as you increase in one variable, you increase in the other variable, so low numbers go with low numbers and high numbers go with high numbers. Using the example mentioned earlier, you may find that as height increases, so does weight — this would be an example of a positive relationship.

With *negative relationships,* as you increase in one variable, you decrease in the other variable, so low numbers on one variable go with high numbers on the other variable. An example of a negative relationship may be that the more depressed you are, the less exercise you do.

You can use the bivariate procedure, which we demonstrate here, whenever you have a positive or negative linear relationship. However, you shouldn't use the bivariate procedure when you have a nonlinear relationship, because the results will be misleading.

Figure 16-2 shows a scatterplot of a nonlinear relationship. As an example of a nonlinear relationship, consider the variables test anxiety (on the *x*-axis) and test performance (on the *y*-axis). People with very little test anxiety may not take a test seriously (they don't study) so they don't perform well; likewise people with a lot of test anxiety may not perform well because the test anxiety didn't allow them to concentrate or even read test questions correctly. However, people with a moderate level of test anxiety should be motivated enough to study, but they don't have too much test anxiety to suffer crippling effects.

Notice that in this example, as we increase in one variable, we increase in the other variable up to a certain point; then as we continue to increase in one variable, we decrease in the other variable. Clearly, there is a relationship between these two variables, but the bivariate procedure would indicate (incorrectly) that there is no relationship between these two variables. For this reason, it's important to always create a scatterplot of any variables you want to correlate so that you don't reach incorrect conclusions.

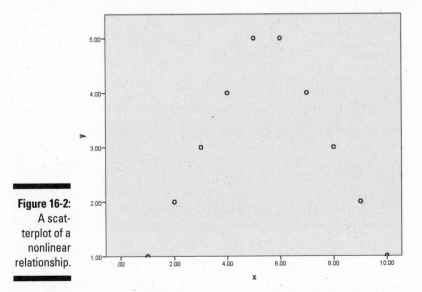

Figure 16-2:
A scatterplot of a nonlinear relationship.

Although a scatterplot visually shows the relationship between two continuous variables, the Pearson correlation coefficient is used to quantify the strength and direction of the relationship between continuous variables. The Pearson correlation coefficient is a measure of the extent to which there is a *linear* (straight line) relationship between two variables. It has values between -1 and $+1$, so that the larger the value, the stronger the correlation. As an example, a correlation of $+1$ indicates that the data fall on a perfect straight line sloping upward (positive relationship), while a correlation of -1 would represent data forming a straight line sloping downward (negative relationship). A correlation of 0 indicates there is no straight-line relationship at all (which is what we would find in Figure 16-2).

To perform a correlation, follow these steps:

1. **From the main menu, choose File ➪ Open ➪ Data and load the** `employee_data.sav` **data file.**

   The file is not in the SPSS installation directory. You have to download it from this book's companion website.

   This file contains the employee information from a bank in the 1960s and has 10 variables and 474 cases.

2. **Choose Analyze ➪ Correlate ➪ Bivariate.**

   The Bivariate Correlations dialog box, shown in Figure 16-3, appears.

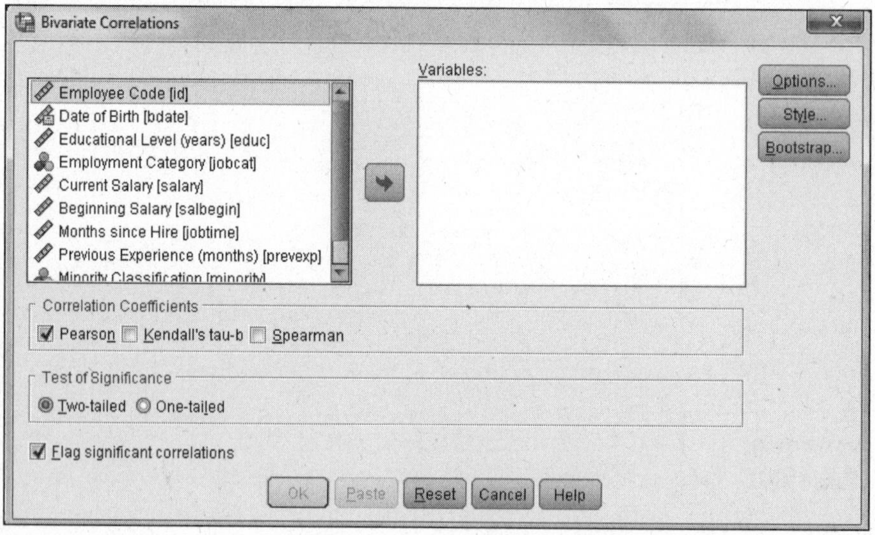

Figure 16-3: The Bivariate Correlations dialog box.

In this example, we want to study whether current salary is related to beginning salary, months on the job, and previous job experience. Notice that there is no designation of dependent and independent variables. Correlations will be calculated on all pairs of variables listed.

3. **Select the variables `salary`, `salbegin`, `jobtime`, and `prevexp`, and place them in the Variables box, as shown in Figure 16-4.**

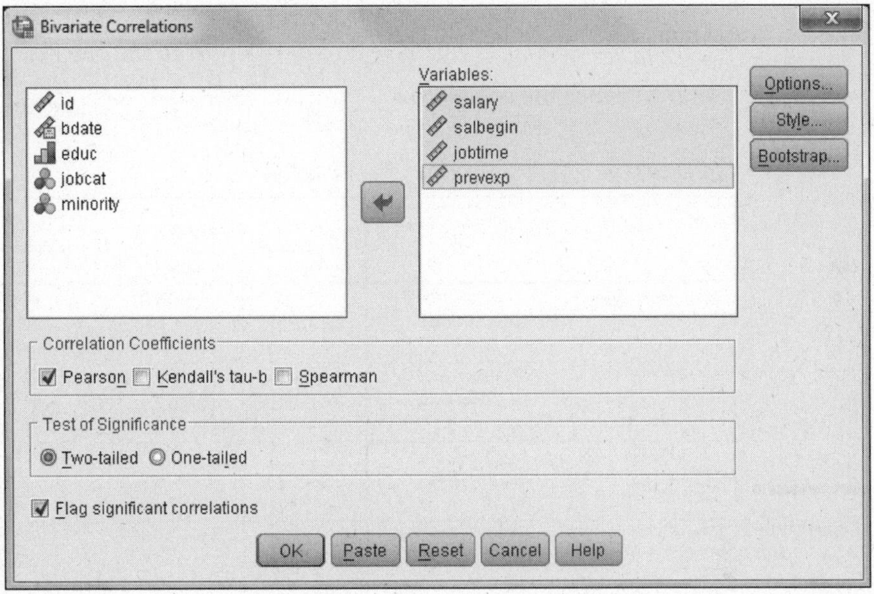

Figure 16-4: The completed Bivariate Correlations dialog box.

You can choose up to three kinds of correlations. The most common form is the Pearson correlation, which is the default. Pearson is used for continuous variables, while Spearman and Kendall's tau-b (less common) are used for nonnormal data or ordinal data, as relationships are evaluated after the original data have been transformed into ranks.

If you want, you can click the Options button and decide what is to be done about missing values and tell SPSS Statistics whether you want to calculate the standard deviations.

4. **Click OK.**

SPSS calculates the correlations between the variables.

Statistical tests are used to determine whether a relationship between two variables is statistically significant. In the case of correlations, we want to test whether the correlation differs from zero (zero indicates no linear association). Figure 16-5 is a standard Correlations table. First, notice that the table is symmetric, so the same information is represented above and below the major diagonal. Also, notice that the correlations in the major diagonal are 1, because these are the correlations of each variable with itself.

The Correlations table provides three pieces of information:

- ✔ The Pearson Correlation, which will range from +1 to –1. The further away from 0, the stronger the relationship.

- ✔ The two-tailed significance level. All correlations with a significance level less than 0.05 will have an asterisk next to the coefficient.

- ✔ N, which is the sample size.

Correlations

		salary Current Salary	salbegin Beginning Salary	jobtime Months since Hire	prevexp Previous Experience (months)
salary Current Salary	Pearson Correlation	1	.880**	.084	-.097*
	Sig. (2-tailed)		.000	.067	.034
	N	474	474	474	474
salbegin Beginning Salary	Pearson Correlation	.880**	1	-.020	.045
	Sig. (2-tailed)	.000		.668	.327
	N	474	474	474	474
jobtime Months since Hire	Pearson Correlation	.084	-.020	1	.003
	Sig. (2-tailed)	.067	.668		.948
	N	474	474	474	474
prevexp Previous Experience (months)	Pearson Correlation	-.097*	.045	.003	1
	Sig. (2-tailed)	.034	.327	.948	
	N	474	474	474	474

Figure 16-5: The Correlations table.

**. Correlation is significant at the 0.01 level (2-tailed).

*. Correlation is significant at the 0.05 level (2-tailed).

In our data, we have a very strong positive correlation (0.880) that is statistically significant between current and beginning salary. Notice that the probability of the null hypothesis being true for this relationship is extremely small (less than 0.01). So, we reject the null hypothesis and conclude that there is a positive, linear relationship between these variables.

The correlations between months on the job and all the other variables were not statistically significant. Surprisingly, we do see that there is a statistically significant negative correlation, although weak (–0.097), between current salary and previous job experience.

Every statistical test has assumptions. The better you meet these assumptions, the more you can trust the results of the test. The Pearson correlation coefficient has three assumptions:

✔ You have continuous variables.

✔ The variables are linearly related.

✔ The variables are normally distributed.

Running the Linear Regression Procedure

Correlations allow you to determine if two continuous variables are linearly related to each other. So, for example, current and beginning salaries are positively related for employees. Regression analysis is about predicting the future (the unknown) based on data collected from the past (the known). Regression allows you to further quantify relationships by developing an equation predicting, for example, current salary based on beginning salary. Linear regression is a statistical technique that is used to predict a continuous dependent variable from one or more continuous independent variables.

When there is a single independent variable, the relationship between the independent variable and dependent variable can be visualized in a scatterplot, as shown in Figure 16-6.

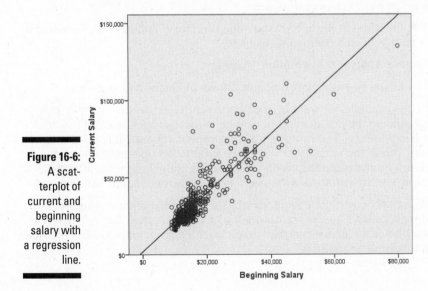

Figure 16-6:
A scatterplot of current and beginning salary with a regression line.

The line superimposed on the scatterplot is the best straight line that describes the relationship. The line has the equation, $y = mx + b$, where, m is the slope (the change in y for a one-unit change in x) and b is the y-intercept (the value of y when x is zero).

In the scatterplot, notice that many points fall near the line, but some are quite a distance from it. For each point, the difference between the value of the dependent variable and the value predicted by the equation (the value on the line) is called the *residual* (also known as the *error*). Points above the line have positive residuals (they were underpredicted), and points below the line have negative residuals (they were overpredicted); a point falling on the line has a residual of zero (a perfect prediction). The regression equation is built so that if you were to add up all the residuals (some will be positive and some will be negative), they would sum to zero.

Overpredictions and underpredictions constitute noise in the model, and noise is normal. All models have some error. A way of thinking about R Square (discussed and defined in the next section) is that this noise is "unexplained variance." R Square, or the signal in our model, is a measure of "explained variance." Add them up and you get the total variance, which we just call variance — the same variance that we use for measures like standard deviation. Conceptually, it makes a lot of sense.

To perform a linear regression, follow these steps:

1. **From the main menu, choose File ➪ Open ➪ Data and load the** `employee_data.sav` **data file.**

   The file is not in the SPSS installation directory. You have to download it from this book's companion website.

2. **Choose Analyze ➪ Regression ➪ Linear.**

   The Linear Regression dialog box, shown in Figure 16-7, appears.

   In this example, we want to predict current salary from beginning salary, months on the job, number of years of education, gender, and previous job experience. You can place the dependent variable in the Dependent box; this is the variable for which we want to set up a prediction equation. You can place the predictor variables in the Independent(s) box; these are the variables we'll use to predict the dependent variable.

   When only one independent variable is taken into account, the procedure is called a *simple regression*. If you use more than one independent variable, it's called a *multiple regression*. All dialog boxes in SPSS provide for multiple regression.

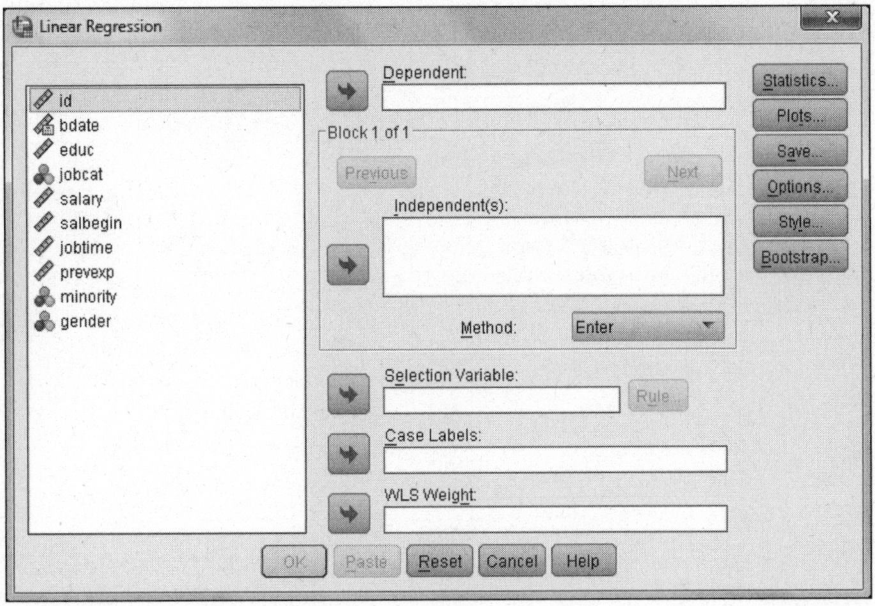

Figure 16-7:
The Linear
Regression
dialog box.

Linear Regression dialog box showing variable list: id, bdate, educ, jobcat, salary, salbegin, jobtime, prevexp, minority, gender. Fields: Dependent, Block 1 of 1 with Previous/Next, Independent(s), Method: Enter, Selection Variable with Rule, Case Labels, WLS Weight. Buttons: Statistics, Plots, Save, Options, Style, Bootstrap, OK, Paste, Reset, Cancel, Help.

3. Select the variable `salary`, and place it in the Dependent box.

4. Select the variables `salbegin`, `jobtime`, `educ`, `gender`, and `prevexp`, and place them in the Independent(s) box, as shown in Figure 16-8.

Note that gender is a dichotomous variable coded 0 for males and 1 for females, but it was added to the regression model. This is because a variable coded as a dichotomy (say, 0 and 1) can technically be considered a continuous variable because a continuous variable assumes that a one-unit change has the same meaning throughout the range of the scale. If a variable's only possible codes are 0 and 1 (or 1 and 2, or whatever), then a one-unit change does mean the same change throughout the scale. Thus, dichotomous variables (for example, gender) can be used as predictor variables in regression. It also permits the use of nominal predictor variables if they're converted into a series of dichotomous variables; this technique is called *dummy coding*.

The last choice we need to make to perform linear regression is that we need to specify which method we want to use. By default, the Enter regression method is used, which means that all independent variables will be entered into the regression equation simultaneously. This method works well when you have a limited number of independent variables or you have a strong rationale for including all your independent variables. However, at times, you may want to select predictors from a larger set of independent variables; in this case, you would request the Stepwise method so that the best predictors from a statistical sense are used.

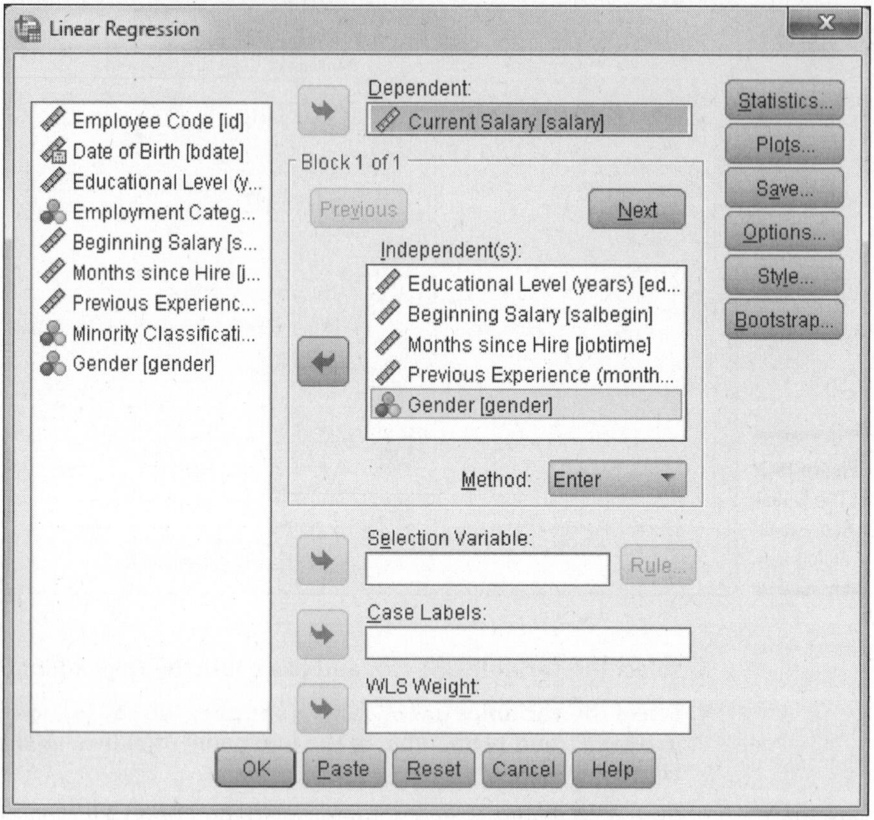

Figure 16-8:
The
completed
Linear
Regression
dialog box.

At this point, you can run the linear regression procedure, but we want to briefly point out the general uses of some of the other dialog boxes:

- The Statistics dialog box has many additional descriptive statistics, as well as statistics that determine variable overlap.

- The Plots dialog box is used to create graphs that allow you to better assess assumptions.

- The Save dialog box adds new variables (predictions, errors) to the data file.

- The Options dialog box controls the criteria when running stepwise regression and choices in handling missing data.

5. Click OK.

SPSS performs linear regression.

Performing regression analysis is the process of looking for predictors and determining how well they predict a future outcome.

The Model Summary table (shown in Figure 16-9) provides several measures of how well the model fits the data. R (which can range from 0 to 1) is the correlation between the dependent measure and the combination of the independent variable(s), so the closer R is to 1, the better the fit. In this example, we have an R of 0.902, which is huge. This is the correlation between the dependent variable and the combination of the five independent variables we're using. You can also think of R as the correlation between the dependent variable and the predicted values.

You may notice that in most stats books, the *r* is lowercase, and for R Square, the *R* is uppercase. Make note of this when you're deciding how to write up your work.

Remember that the ultimate goal of linear regression is to create a prediction equation so we can predict future values. The value of the equation is linked to how well it actually describes or fits the data, so part of the regression output includes fit measures. To quantify the extent to which the straight-line equation fits the data, the fit measure, R Square, was developed. R Square (which can range from 0 to 1) is the correlation coefficient squared. It can be interpreted as the proportion of variance of the dependent measure that can be predicted from the combination of independent variable(s). In this example, we have an R Square of 0.814, which is huge. This value tells us that our combination of five predictions can explain about 81% of the variation in the dependent variable, current salary. See Table 16-1 for more context of how large a R Square of 81% is.

It's reasonable for folks to disagree a bit about what constitutes a big correlation. For instance, if you're a chemist or a physicist, correlations would be expected to be very high because physical objects follow natural laws quite consistently. When human behavior is involved, however, even correlations in the 0.3 to 0.5 range, which would correspond to an R Square of 10% to 25%, are quite high. Some research would report correlations in the 0.1 range, but when you square that, you realize that it's really pretty low.

Table 16-1	Some R Value Ranges and Their Equivalent R Square Value Ranges	
	r	*R Square*
Noteworthy	Greater than 0.3	Greater than 9% to 10%
Large	Greater than 0.5	Greater than 25%
Very large	Greater than 0.7	Greater than 49% to 50%

Adjusted R Square represents a technical improvement over R Square in that it explicitly adjusts for the number of predictor variables relative to the sample size. If Adjusted R Square and R Square differ dramatically, it's a sign either that you have too many predictors or that your sample size is too small. In our situation, Adjusted R Square has a value of 0.812, which is very similar to the R Square value of 0.814; so we aren't capitalizing on chance by having too many predictors relative to the sample size.

The Standard Error of the Estimate provides an estimate (in the scale of the dependent variable) of how much variation remains to be accounted for after the prediction equation has been fit to the data.

Model Summary

Model	R	R Square	Adjusted R Square	Std. Error of the Estimate
1	.902[a]	.814	.812	$7,410.457

a. Predictors: (Constant), gender Gender, jobtime Months since Hire, prevexp Previous Experience (months), salbegin Beginning Salary, educ Educational Level (years)

Figure 16-9: The Model Summary table.

While the fit measures in the Model Summary table indicate how well you can expect to predict the dependent variable, they don't tell us whether there is a statistically significant relationship between the dependent and the combination of independent variable(s). The ANOVA table is used to determine whether a statistically significant relationship between the dependent variable and the combination of independent variables — that is, if the correlation between dependent and independent variables differs from zero (zero indicates no linear association). The Sig. column provides the probability that the null hypothesis is true — that is, no relationship between the independent(s) and dependent variables. As shown in Figure 16-10, in our case, the probability of the null hypothesis being correct is extremely small (less than 0.05), so the null hypothesis has to be rejected, and the conclusion is that there is a linear relationship between these variables.

ANOVA[a]

Model		Sum of Squares	df	Mean Square	F	Sig.
1	Regression	1.122E+11	5	2.244E+10	408.692	.000[b]
	Residual	2.570E+10	468	54914875.01		
	Total	1.379E+11	473			

a. Dependent Variable: salary Current Salary

b. Predictors: (Constant), gender Gender, jobtime Months since Hire, prevexp Previous Experience (months), salbegin Beginning Salary, educ Educational Level (years)

Figure 16-10: The ANOVA table.

Because the results from the ANOVA table were statistically significant, we turn next to the Coefficients table. If the results from the ANOVA table were not statistically significant, we would conclude that there was no relationship between the dependent variable and the combination of the predictors, so there would be no reason to continue investigating the results. Because we do have a statistically significant model, however, we want to determine which predictors are statistically significant (it could be all of them or maybe just one). We also want to see our prediction equation, as well as determine which predictors are the most important. To answer these questions, we turn to the Coefficients table (shown in Figure 16-11).

Coefficients[a]

Model		Unstandardized Coefficients		Standardized Coefficients	t	Sig.
		B	Std. Error	Beta		
1	(Constant)	-12550.032	3474.744		-3.612	.000
	salbegin Beginning Salary	1.723	.061	.794	28.472	.000
	jobtime Months since Hire	154.536	34.085	.091	4.534	.000
	prevexp Previous Experience (months)	-19.436	3.583	-.119	-5.424	.000
	educ Educational Level (years)	593.031	166.630	.100	3.559	.000
	gender Gender	-2232.917	792.078	-.065	-2.819	.005

a. Dependent Variable: salary Current Salary

Figure 16-11: The Coefficients table.

Linear regression takes into consideration the effect one or more independent variables have on the dependent variable. In the Coefficients table, the independent variables appear in the order they were listed in the Linear Regression dialog box, not in order of importance. The B coefficients are important for both prediction and interpretive purposes; however, analysts usually look first to the t test at the end of each row to determine which independent variables are significantly related to the outcome variable. Because five variables are in the equation, we're testing if there is a linear relationship between each independent variable and the dependent variable after adjusting for the effects of the four other independent variables. Looking at the significance values, we see that all five of the predictors are statistically significant, so we need to retain all five of the predictors.

The first column of the Coefficients table contains a list of the independent variables plus the *constant* (the intercept where the regression line crosses the y-axis). The intercept is the value of the dependent variable when the independent variable is 0.

The B column shows you how a one-unit change in an independent variable impacts the dependent variable. For example, notice that for each additional year of education completed, the expected increase in current salary is $593.03. The variable months hired has a B coefficient of $154.54, so each additional month increases current salary by $154.54. Whereas the variable previous experience has a B coefficient of –$19.44, each additional month decreases current salary by –$19.44.

The variable gender has a B coefficient of about –$2,232.92. This means that a one-unit change in gender (which means moving from male to female), is associated with a drop in current salary of –$2,232.92. Finally, the variable beginning salary has a B coefficient of $1.72, so each additional dollar increases current salary by $1.72.

The B column also contains the regression coefficients you would use in a prediction equation. In this example, current salary can be predicted with the following equation:

Current Salary = –12550 + (1.7)(Beginning Salary) + (154.5)(Months Hired) + (–19.4)(Previous Experience) + (593)(Years of Education) + (–2232.9) (Gender)

The Std. Error column contains standard errors of the regression coefficients. The standard errors can be used to create a 95% confidence interval around the B coefficients.

If we simply look at the B coefficients you may think that gender is the most important variable. However, the magnitude of the B coefficient is influenced by amount of variation of the independent variable. The Beta coefficients explicitly adjust for such variation differences in the independent variables. Linear regression takes into account which independent variables have more impact than others.

Betas are standardized regression coefficients and are used to judge the relative importance of each of the independent variables. The values range between –1 and +1, so that the larger the value, the greater the importance of the variable. In our example, the most important predictor is beginning salary, followed by previous experience, and then education level.

Every statistical test has assumptions. The better you meet these assumptions, the more you can trust the results of the test. Linear regression assumes the following:

✔ You have continuous variables.

✔ The variables are linearly related.

✔ The variables are normally distributed.

✔ The errors are independent of the predicted values.

✔ The independent variables are not highly related to each other.

Part VI

Making SPSS Your Own: Settings, Templates, and Looks

 See how to create your own TableLook in a free article at www.dummies.com/extras/spss.

In this part . . .

- Make SPSS your own with customized features and appearance.

- Learn all about graph editing and how to save your customizations for the future.

- Find out how to edit tables and where to find table looks in SPSS.

Chapter 17

Changing Settings

. . . .

In This Chapter

▶ Understanding your configuration options

▶ Modifying the default settings

. . . .

*O*ver time, you'll find that you want to configure your system to work in ways that are different from the defaults. SPSS has lots of options that you can set to do just that. If you're new to this and you've just started looking at the software, you probably don't want to change any options just yet, but you need some idea of the possibilities it offers. Later, when you absolutely, positively have to make some sort of change, you'll know where to go to do it.

With the Data Editor window on the screen (see Figure 17-1), choose Edit ➪ Options to display the Options window. You can set any and all possible options in this window. At the top of the window are some tabs; each tab contains a different collection of options. In this chapter, we walk you through changing these options.

Sometimes a change in configuration has an immediate effect, and sometimes it doesn't. For example, if you change the way values are labeled in a report that's already onscreen, nothing happens because the report has already been constructed. You have to run the report-generating software again — so it builds a new report — to have the changes take effect.

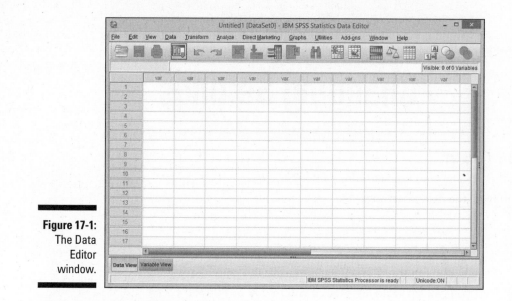

Figure 17-1:
The Data
Editor
window.

General Options

The first tab in the Options window, the General tab (shown in Figure 17-2), displays options that don't fit into any of the other categories. The options displayed on the General tab are as follows:

- ✔ **Variable Lists:** Lists of variables in your output can be identified by their labels or their names. (Think of names as short titles and labels as long titles.) You can have your data items, by default, tagged by one or the other as they appear in variable lists. Also, you can have your data appear in alphabetical order by the titles you use for individual items, or simply by the order in which the data appears in the file. File order usually makes more sense.

- ✔ **Roles:** When you select some actions, variables of the types that can play certain roles in the processing to follow can be preselected for you if you have the first option (Use Predefined Roles) selected. If you have the other option (Use Custom Assignments) selected, you'll be required to choose all the variables yourself.

- ✔ **Output:** These options control the appearance of tables and graphs:

    - **No Scientific Notation for Small Numbers:** Suppresses scientific notation for small numbers. For example, 12 appears as 12 instead of 1.2e1, which is a little harder to read. SPSS doesn't say exactly what it considers to be a small number.

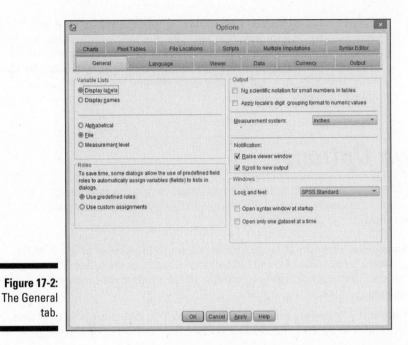

Figure 17-2:
The General tab.

- **Apply Locale's Digit Grouping Format to Numeric Values:** Applies the current locale's digit grouping format to numeric values in pivot tables and charts as well as in the Data Editor.

- **Measurement System:** This is the units used to specify the margins between table cells, the width of cells, and the spacing between printed characters. You can use inches, centimeters, or points. (A point is $1/72$ of an inch.)

✔ **Notification:** This is the method the software uses to notify you when the results of a calculation are available. With the Raise Viewer Window option, the display window opens automatically. With the Scroll to New Output option, the window scrolls and exposes the location of the new data.

✔ **Windows:** These are cosmetic options. You can choose how you like the dialog boxes of SPSS to appear.

- **Look and Feel:** Your choices are SPSS Standard, SPSS Classic, and Windows.

- **Open Syntax Window at Startup:** Makes SPSS begin with the Syntax window instead of the Data Editor window. Choose this option if you use the scripting language more often than the windowing interface to enter data and run your predefined procedures.

- **Open Only One Dataset at a Time:** Whenever you open a new dataset, the new information appears in a new window and any that are already open are closed. With this option selected, the already open dataset is closed when the new one is opened. By the way, this does not apply to datasets opened inside a Syntax language process.

Language Options

The Language tab (shown in Figure 17-3) displays options for changing language and character encoding. The options displayed on the Language tab are as follows:

- ✔ **Language:** The output language can be set to any one of about a dozen choices, and it determines the language used to output files. You may have to use Unicode mode to handle all the characters in some languages.

- ✔ **Character Encoding for Data and Syntax:** You can choose to read and write files in Unicode mode, but you shouldn't unless you have a good

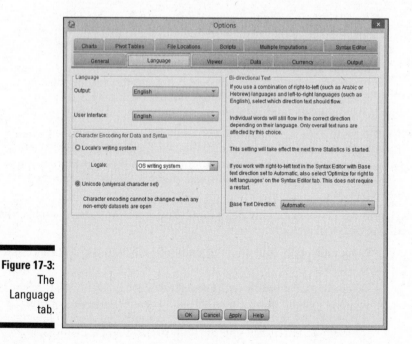

Figure 17-3:
The Language tab.

reason to do so. The files are read and written in UTF-8 format. If you write a Unicode file, you need to be sure that the software that reads it understands that format. When you read a file in Unicode mode, it's much larger in memory than it would be otherwise.

✔ **Bi-Directional Text:** Here, you can select which direction text should flow in the software. You can choose to use left-to-right languages (such as English) or right-to-left languages (such as Arabic and Hebrew). Make sure to restart SPSS to see your selected changes.

Viewer Options

Output from SPSS is formatted for viewing with either the draft viewer or the regular viewer. SPSS thinks in terms of a printed page, but the same layouts are used for displaying data onscreen. The options you can set for the regular viewer can be accessed with the Viewer tab, shown in Figure 17-4. The options in the Viewer tab are as follows:

✔ **Initial Output State:** Determines which items are displayed each time you run a procedure. You choose an item either by selecting its name (Log, Warnings, Notes, Title, and so on) from the drop-down list or by selecting its icon. Then you can select whether you want it to appear or remain hidden, how you want its text justified (Align Left, Center, or Align Right), and whether the information occurrence should be included as part of the log (Display Commands in the Log).

✔ **Title:** Here you choose the font used for main output titles. It appears at the top of the first page of a report.

✔ **Page Title:** Choose the font used for the title appearing at the top of subsequent pages of a report.

✔ **Text Output:** Determines the font used for the text of your report and the labels on graphs and tables. The font size also affects the page width and length because the sizes are measured by counting characters.

Some fonts have variable-width characters, which throws off the alignment of your columns. If you want everything to align in neat columns all the time, use the monospaced font.

✔ **Default Page Setup:** Here you see the default settings for the Orientation and Margins. You can choose Portrait or Landscape to change the orientation settings, as well as set your own margins.

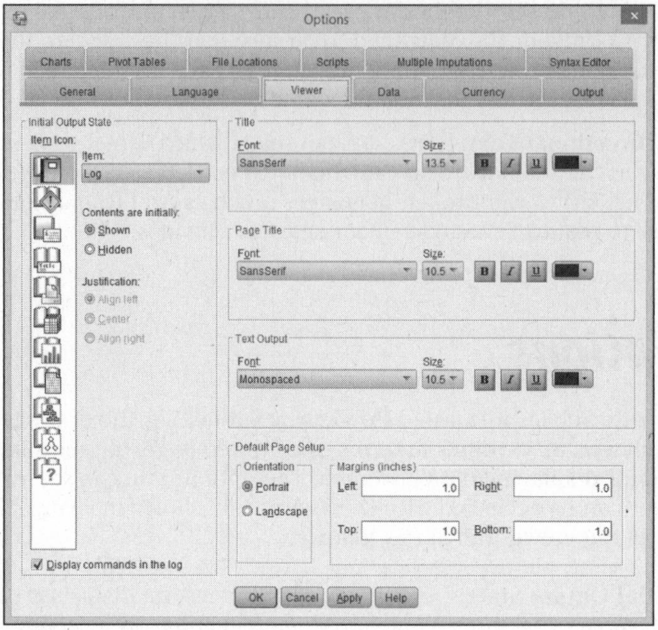

Figure 17-4:
The Viewer
tab.

Data Options

The Data tab, shown in Figure 17-5, can be used to specify how SPSS handles a few special numeric situations. The options in the Data tab are as follows:

- **Transformation and Merge Options:** Determines when — not how — results are calculated. You can have SPSS perform calculations immediately, or you can have it wait until it needs the number for something (either another calculation or a displayed value).

- **Display Format for New Numeric Variables:** Determines how many digits are used in the display of values and how many digits are to the right of the decimal. Width is the total number of characters, including the decimal point. The Decimal Places setting determines the number of digits that appear to the right of the decimal point. If the number of places to the right is too small, values are rounded to fit. If the number of places is too large, values are put in scientific notation.

- **Random Number Generator:** Ever since a need for random numbers was discovered, generating them on a computer has been a problem — because computers naturally do things in a *non*-random way. SPSS offers you two ways to do it: the old way and the new way. If you'd rather not generate your random numbers the same way you did in older versions of SPSS (version 12 and earlier), use the long period Marsenne Twister.

- ✔ **Set Century Range for 2-Digit Years:** A solution to the Y2K problem. You thought that was all over, right? It is, but the solutions are still with us — and this is one of them: You put in two four-digit years here, and any two-digit value that you supply to identify a year is assumed to be between the two years you specify. This is mostly for old data. If you always use four digits for years in your data, this setting is never used.

- ✔ **Customize Variable View:** Allows you to determine which variable attributes are displayed and in what order they're displayed in the Variable View tab of the Data Editor window.

- ✔ **Change Dictionary:** Allows you to determine which dictionary is used to check the spelling on the Variable View tab.

- ✔ **Assigning Measurement Level:** When SPSS reads numeric data, it counts the number of unique values assigned to a variable and uses the count to determine whether the variable is nominal or scalar. The count you enter here determines the threshold used to make that determination.

- ✔ **Rounding and Truncation of Numeric Values:** This setting determines the threshold for rounding numbers. SPSS does the calculation in base two, so the count is a number of bits. *Fuzz bits* refers to a count of the number of bits to be considered. The setting is specified to 6 at install time, which should be sufficient for most applications.

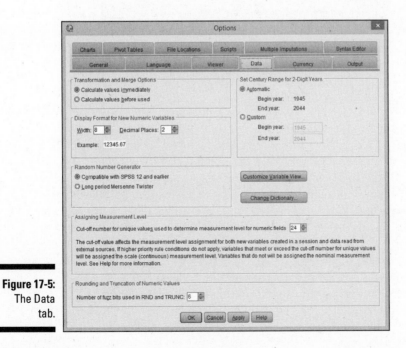

Figure 17-5:
The Data tab.

Currency Options

Different parts of the world use different symbols and formats when writing about currency. The Currency tab (shown in Figure 17-6) lets you specify the display format of your currency. Here are the options on the Currency tab:

- **Custom Output Formats:** Here you can specify the default format for presenting currency values. The five formats have the unlikely names CCA, CCB, CCC, CCD, and CCE. Those are the only ones you can have, but that has to be enough for anybody. The calculations are always performed the same way — the differences are in the display. You can set the display configuration for each one to be anything you'd like (dollars, euros, yen, and so on), and then switch among them as often as you want.

- **Sample Output:** Displays the printed format of positive and negative currency values. As you switch from one currency selection to another, and as you change the formatting of any of them, the sample displays examples of the format.

- **All Values:** Specifies the characters that appear onscreen to identify the currency, at the front or back of all values. Such characters include the British pound sign and the cent mark.

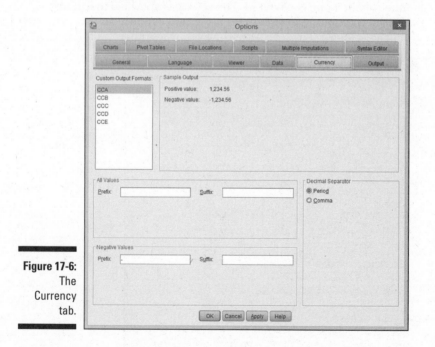

Figure 17-6:
The Currency tab.

✔ **Decimal Separator:** Many currency notations use commas instead of periods to denote the fractional portion of the amount.

✔ **Negative Values:** Specifies the characters placed in front and in back of negative values. For example, some folks like to use < and > to surround negative money values.

Output Options

Every variable can be identified in two ways: by a label and by a name. In your output, you can specify to have variables identified by one or the other or both. You configure output labeling on the Output tab, shown in Figure 17-7. Following are the options in the Output tab:

✔ **Outline Labeling:** The text used to identify the parts of charts and graphs.

✔ **One Click Descriptives:** Allows you to suppress or place a limit on the columns in a table output.

✔ **Output Display:** Allows you to choose if you want to view the output using the Model Viewer or using pivot tables and charts.

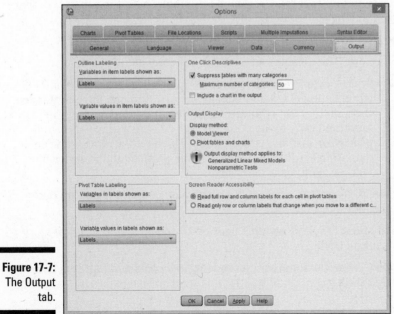

Figure 17-7:
The Output tab.

✔ **Pivot Table Labeling:** The text used to identify the rows and columns of tables.

✔ **Screen Reader Accessibility:** Allows you to change the accessibility of rows and columns in an output.

Longer labels can be descriptive and make your data easier to determine, but they can also screw up some formats.

Chart Options

The default appearance of charts is determined by the settings on the Charts tab, shown in Figure 17-8. The options in the Chart tab are as follows:

✔ **Chart Template:** A file that contains a set of starter settings that you can use for designing a new chart. When you create a new chart, it can use the settings in this configuration window, or it can use this file. You can select any file to be your default starting template. It's easy to create a chart template: Simply create a chart that has all the configuration settings you like — and save it so you can use it as the template file.

✔ **Chart Aspect Ratio:** The ratio of the width to the height of the produced charts, initially set to 1.25. Which ratio looks better is a matter of opinion; you'll have to experiment.

✔ **Current Settings:** This section offers two drop-down menus:

  • **Font:** The default font for the text in any chart you design.

  • **Style Cycle Preference:** How SPSS chooses the styles and colors when laying out data items in a chart. You can have SPSS cycle through just the colors so each item included in the graph is identified only by its color. If you're using a black-and-white printer or display, choose Cycle Through Patterns Only; each data item is identified by a graphic pattern of line styles and marker symbols.

✔ **Frame:** Determines whether charts display an inner frame, an outer frame, both, or neither.

✔ **Grid Lines:** Displays dividing lines on the scale axis, on the category axis, or on both.

✔ **Style Cycles:** Customizes the sequence of colors and patterns to be cycled through.

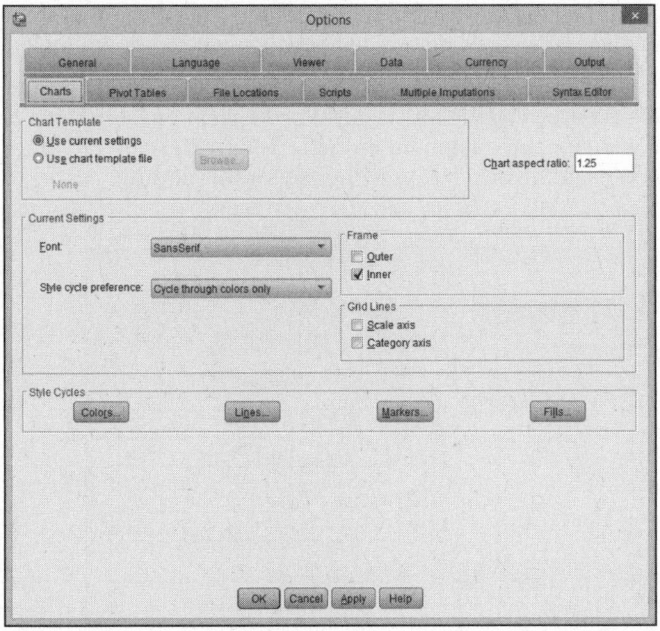

Figure 17-8:
The Charts
tab.

Pivot Tables Options

The tabular output format of SPSS is the *pivot table*. The Pivot Tables tab (shown in Figure 17-9) is used to set display options for the tables. The options on the Pivot Tables tab are as follows:

- ✔ **TableLook:** A file that contains your standard pivot table and determines the initial appearance of any new tables you create. Several such files come with the system and are listed in the dialog box. You can also create your own file by choosing TableLook from the menu in the Pivot Table Editor window. The Set TableLook Directory button sets the currently displayed directory as the one in which your new table files are stored. You can choose any directory you like; clicking this button causes your chosen directory to appear in this window by default.

- ✔ **Column Widths:** Controls the way SPSS adjusts column widths in pivot tables. You can adjust them according to the width of the labels or according to the width of the data and labels, whichever is wider.

- ✔ **Table Comment:** Automatically includes comments for each table.

- ✔ **Sample:** Shows an example of how each TableLook will render.

- ✔ **Table Rendering:** Using this option allows compatibility with tables that were rendered with SPSS prior to version 20. For version 20 and higher, all tables have full support for pivoting and editing.

✔ **Default Editing Mode:** When you double-click a pivot table, you can edit it in place or a separate edit window is opened, depending on this setting.

✔ **Copying Wide Tables to the Clipboard in Rich Text Format:** When a table is copied to the Microsoft Word format or Rich Text Format, tables too wide for the document are wrapped to fit, scaled to fit, or left as they are, depending on what you choose for this setting.

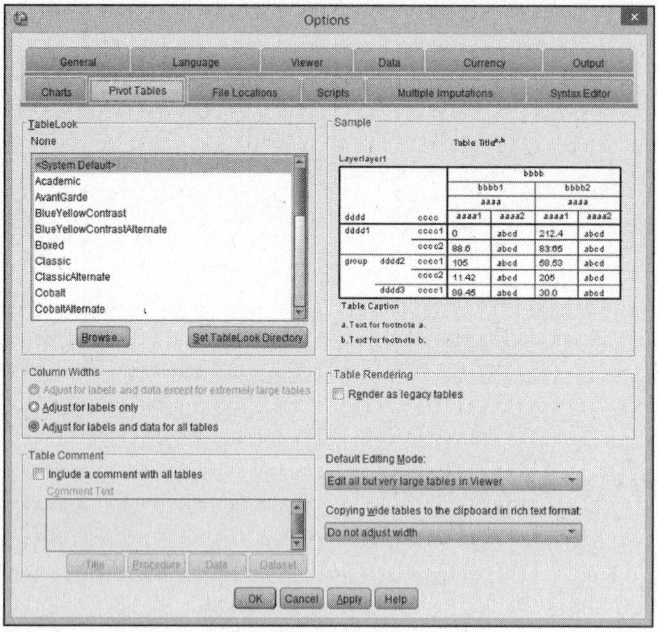

Figure 17-9:
The Pivot
Tables tab.

File Locations Options

The options on the File Locations tab (shown in Figure 17-10) specify the locations of the files opened for input and output. The options for the file locations are as follows:

✔ **Startup Folders for Open and Save Dialogs:** The startup folders are the names of the directories that initially appear in the Save and Open dialog boxes when you read or write data files. Optionally, you can select to simply use the last directory used to read or write a file.

✔ **Session Journal:** You have the option to configure a journal file to receive a copy of every Syntax language command, whether it comes from a script or from a user entering instructions through a dialog box.

✔ **Temporary Folder:** You can specify the name of the directory where SPSS creates its temporary working files.

✔ **Number of Recently Used Files to List:** The most recently read or written files are listed in the File menu. This option specifies how many are listed.

✔ **Python Location:** If you installed Python during the installation of SPSS Statistics 23, you don't need to worry about this. If you didn't, you can specify the name of the directory where you have a version of Python installed.

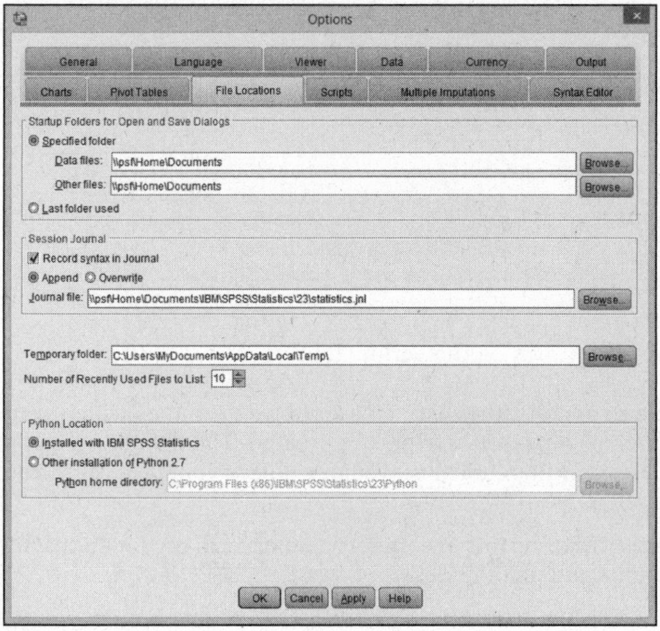

Figure 17-10:
The File
Locations
tab.

Scripts Options

The Scripts tab (shown in Figure 17-11) is used to determine some fundamental defaults about scripts.

Don't mess with any of these options until you've been writing scripts for a bit and know what you're doing — a single change here can affect the execution of a number of scripts.

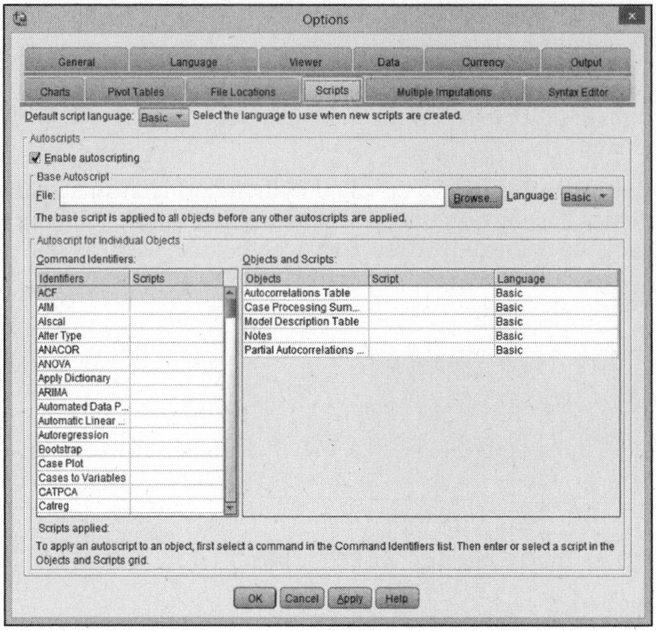

Figure 17-11:
The Scripts
tab.

Here are the options on the Scripts tab:

✔ **Default Script Language:** This setting determines which script editor is launched when new scripts are created. The default script language is Basic. No other choice is available unless you have the Python add-on installed.

✔ **Autoscripts:** Scripts are used to automate many functions, including customizing pivot tables.

- **Enable Autoscripting:** If you choose this option, you enable the autoscripting feature.

- **Base Autoscript:** A script, if it's stored in the file you name here, defines a global procedure that runs automatically when you create an object. It always runs before any other autoscript for that object. The choice of languages for it can be either Basic or Python (and then only if the Python add-on is installed).

- **Autoscript for Individual Objects:** By associating a type of object with an autoscript, you can make an autoscript execute when an object of that specific type is created. To associate an autoscript with an object type, first select the command that generates an object of the desired type (these commands appear in the Identifiers column on the left). On the right, the Objects column displays the types of objects that your chosen identifier command will generate. In the script cell to the right of the object type you want to tag, enter the path name of the file containing the auto-script. Alternatively, you can click the ellipsis button that appears in the cell and browse for a script file. When you've chosen the file you want, click OK or Apply to make the association.

To delete an autoscript association, in the Script column on the right, select the name of the script file you want to disassociate, and then delete it. Select some other cell to make sure your deletion has been accepted, and then click OK or Apply.

Multiple Imputations Options

SPSS keeps track of which data has been entered and which has been *imputed* (assumed). The imputation process is that of calculating what the values of your missing data *would be*. You can set multiple imputation options using the Multiple Imputations tab (shown in Figure 17-12). Here are the options on the Multiple Imputations tab.

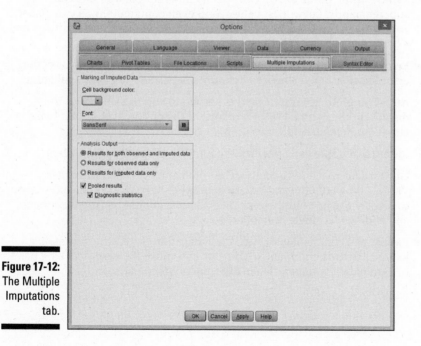

Figure 17-12:
The Multiple Imputations tab.

- **Marking of Imputed Data:** You can change the appearance of imputed data on the Data View tab of the Data Editor window. You can highlight it by changing the background color of the cell in which it's displayed, and by using a different font to write its values.

- **Analysis Output:** You can choose to do analytical calculations using imputed data, without using it, or both ways. Also, you can set the imputation process to pool previously imputed data for further imputation. We suggest leaving this setting alone for now — it takes a mathematician to figure it all out. Some analysis procedures can produce separate analysis results using only imputed data — you can choose to generate output from such pooled data.

Syntax Editor Options

The editor of the Syntax Command language is capable of recognizing various language parts and highlighting them for you. The Syntax Editor tab (shown in Figure 17-13) has the following options:

- **Syntax Color Coding:** You can specify different colors for commands, subcommands, keywords, values, and even comments. You can also specify each one as bold, italic, or underlined. A single switch turns on all coloring and highlighting.

- **Error Color Coding:** You can specify the font style and the color coding of error information. A single switch turns on all coloring and highlighting.

- **Auto-Complete Settings:** Use this setting to suppress or allow the display, in the Syntax Editor window, of the option button that turns auto-complete on or off.

- **Indent Size (Spaces):** Specifies the number of spaces in an indent.

- **Gutter:** The space to the left of the commands is called the *gutter.* Various types of information are displayed there. You can use the gutter to display the line numbers or the span of a command (the beginning and ending of a single command).

- **Panes:** You can display or hide the navigation pane, which contains a list of all Syntax commands. You can also cause the error-tracking pane to automatically appear when SPSS encounters an error.

✔ **Optimize for Right to Left Languages:** You have to select this option when you're working in a language that reads right to left (such as Hebrew or Arabic).

✔ **Paste Syntax from Dialogs:** Specifies the position at which Syntax is inserted into the Syntax window when pasting Syntax from a dialog.

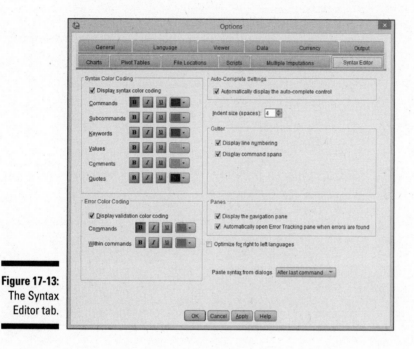

Figure 17-13:
The Syntax
Editor tab.

Chapter 18

Editing Charts and Chart Templates

In Chapter 12, you see how to make simple graphs and fancy graphs. This chapter explains how to edit these graphs to create a better visual representation of the information in the chart. This chapter also helps you learn how to showcase the most important information to your audience.

Although this chapter doesn't cover every type of edit you can make to a chart, you can use the steps here as a baseline to explore additional edit options. When you get the basic idea of how to edit graphs using the Chart Builder, you can continue to explore making edits on your own.

Changing and Editing Axes

An axis on a chart or graph acts as a reference that gives information on how variables are related. In this section, we explain how to edit an axis to help better represent the information in a graph.

Changing the axis range

In Chapter 12, you create a differenced area graph (refer to Figure 12-17). As you can see in the graph, there is empty space between the bottom of the graph and where the information begins along the y-axis. Here we explain how to edit the y-axis so that the range of the axis begins at 50. (Refer to Chapter 12 if you need to re-create the graph.)

The following steps show you how to edit Figure 12-17:

1. **In the SPSS Statistics Viewer window, navigate to the differenced area graph and double-click the chart.**

   The Chart Editor window appears (see Figure 18-1).

2. **Click the Y button at the top of the menu bar.**

   The Properties dialog box appears.

3. **Click the Scale tab (see Figure 18-2).**

4. **In the Range section, click the Auto check box next to Minimum, and change the number in the Custom box to 50.**

5. **Click Apply.**

   The chart in Figure 18-3 appears.

   As you can see, the *y*-axis in this new figure is changed to eliminate unnecessary space in the graph.

Editing graphs like this allows you to create clearer graphs and improve how the information is visualized.

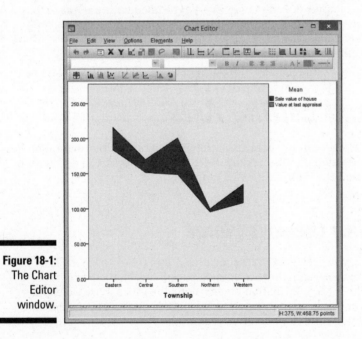

Figure 18-1:
The Chart
Editor
window.

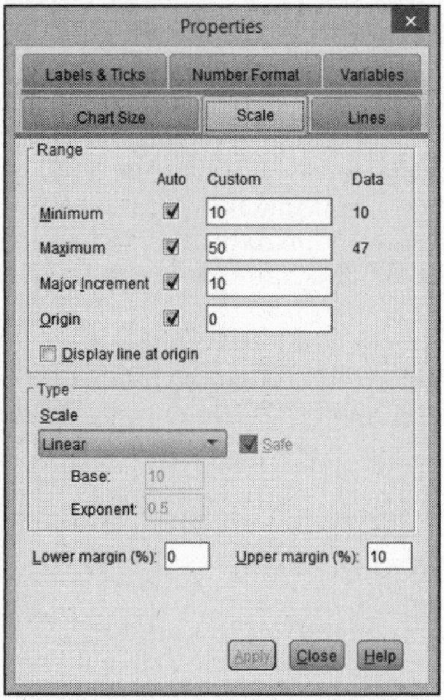

Figure 18-2:
The Scale tab of the Properties dialog box.

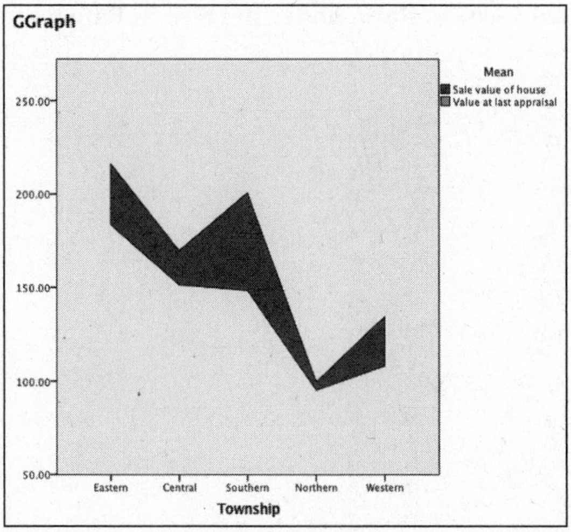

Figure 18-3:
The differenced area graph with the edited *y*-axis.

Scaling the axis range

No plot is simpler to produce than the *dot plot*. It has only one dimension. Although SPSS does an excellent job of grouping the information, it doesn't create a visually scaled chart like the one shown in Figure 12-2. Without scaling, your chart may fail to convey your information with the impact that you want. For example, notice that when you build the dot plot in Chapter 12, the information is scrunched, with too much white space at the top of the graph.

To be able to make this chart more readable, we'll change how SPSS scales the graph. The following steps guide you through the process of editing the dot plot from Chapter 12. (Refer to Chapter 12 if you need to re-create the graph.)

1. **In the SPSS Statistics Viewer window, navigate to the dot plot and double-click the chart.**

   The Chart Editor window appears.

2. **Double-click the graph.**

   The Properties dialog box appears.

3. **Click the Chart Size tab (see Figure 18-4).**

4. **Look for Maintain Aspect Ratio under the Size in Points section.**

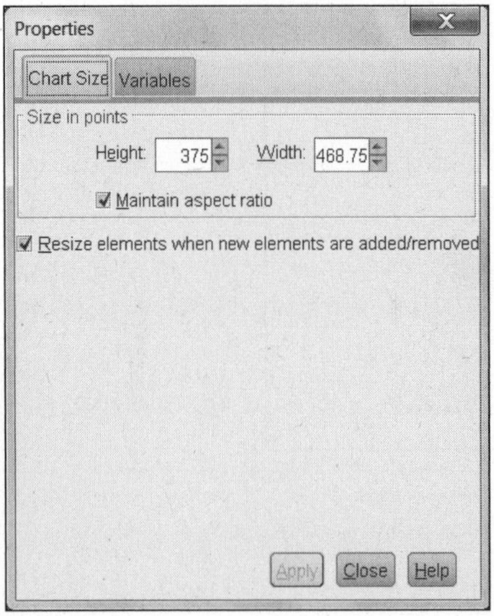

Figure 18-4:
The Chart
Size tab
of the
Properties
dialog box.

5. **Uncheck the Maintain Aspect Ratio button and change the Height to 150.**

6. **Click Apply.**

The chart shown in Figure 18-5 appears.

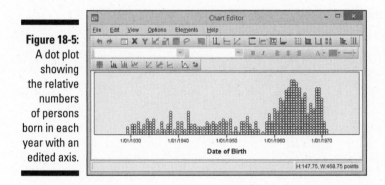

Figure 18-5:
A dot plot showing the relative numbers of persons born in each year with an edited axis.

Changing Style: Lines and Symbols

When creating line graphs and scatterplots, it can be hard to differentiate between lines and data points on a graph, especially in black and white. In this section, we tell you how to change the style of data within your graphs to make it easier to visualize the information.

Editing chart lines

In Chapter 12, we explain how to make a chart with multiple lines (refer to Figure 12-9). In order to help the viewer distinguish the various lines, it's important to modify the way they appear. The following steps can be used to edit the chart for clarity. (Refer to Chapter 12 if you need to re-create the chart.)

1. **In the SPSS Statistics Viewer window, navigate to the multiple line chart and double-click the chart.**

The Chart Editor window appears.

2. **Triple-click one of the lines in the graph.**

Make sure that only one line is selected. The Properties dialog box appears.

3. **Click the Lines tab.**

4. **In the Lines section, from the Style drop-down list (see Figure 18-6), choose a new line style.**

5. **Click Apply.**

   The style you chose is applied to the line.

6. **Repeat the process for the other lines.**

   The line chart in Figure 18-7 appears.

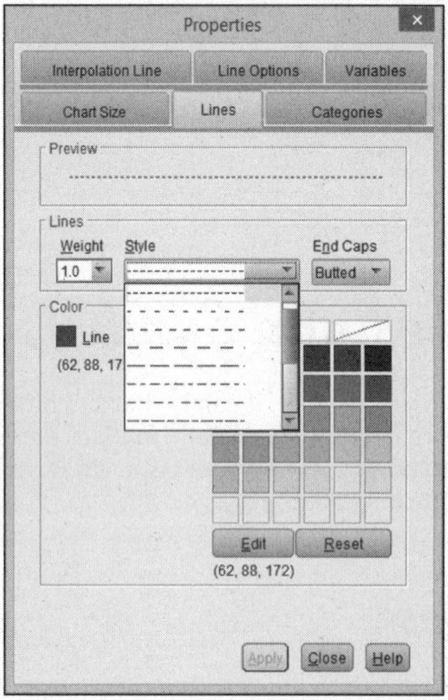

Figure 18-6: The Style drop-down list in the Properties window.

Editing data points

Colored scatterplots such as the one in Figure 12-12 are a great way to visualize information. In situations where color scatterplots are not an option (for example, you don't have a color printer), you need to change the style of the data points so the viewer can distinguish between the datasets.

The following steps can be used to edit the chart and allow a stronger delineation between data sets in a chart. (Refer to Chapter 12 if you need to re-create the scatterplot.)

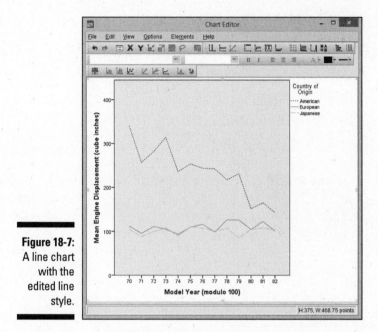

Figure 18-7:
A line chart
with the
edited line
style.

1. **In the SPSS Statistics Viewer window, navigate to the scatterplot and double-click the chart.**

   The Chart Editor window appears.

2. **Double-click one of the circles in the graph.**

   Make sure that only one set of data is selected.

   The Properties dialog box appears.

3. **Click the Marker tab.**

4. **In the Marker section, from the Type drop-down list, choose a new marker type (see Figure 18-8).**

5. **Click Apply.**

   The style you selected is applied to the graph.

6. **Repeat the process for the other datasets.**

   The chart shown in Figure 18-9 appears.

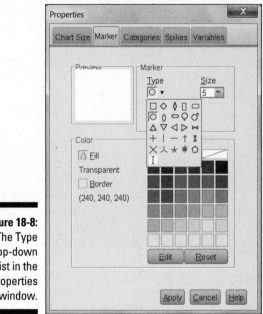

Figure 18-8:
The Type
drop-down
list in the
Properties
window.

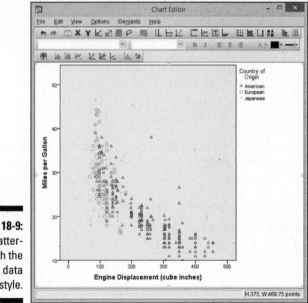

Figure 18-9:
A scatter-
plot with the
edited data
point style.

Applying Templates

Templates are a great way to improve how your graph looks and feels. To save you from having to create templates yourself, SPSS comes with a set of templates for charts and graphs that you can apply to make your graphic stand out. In this section, we discuss how to apply these templates to your graph.

In Figure 12-9 of Chapter 12, we created a chart with multiple lines. Here we apply a grayscale template to add effect to a chart. The following steps show how to apply a template to a chart. (Refer to Chapter 12 if you need to re-create the chart.)

1. **In the SPSS Statistics Viewer window, navigate to the line chart and double-click the chart.**

   The Chart Editor window appears.

2. **Choose File⇨Apply Chart Template.**

   The Apply Template dialog box appears.

3. **Click Greyscale.sgt.**

4. **Click Open.**

   The chart shown in Figure 18-10 appears.

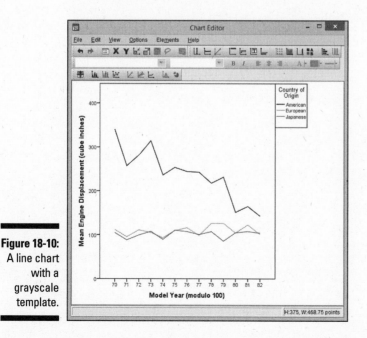

Figure 18-10:
A line chart with a grayscale template.

Chapter 19

Editing Tables

. . . .

In This Chapter

▶ Changing the appearance of tables with TableLooks

▶ Using the Style Output feature to highlight certain cells in your output

▶ Transforming tables using pivoting trays

. . . .

SPSS pivot tables are pretty cool. In fact, in 1997, two years after they came out, SPSS pivot tables were added to the Smithsonian Institutes Permanent Research Collection of Information Technology. Virtually all SPSS tabular output comes in the form of a pivot table. It isn't the product of a special menu. FREQUENCIES tables, DESCRIPTIVES tables, and just about every other command make them.

In this chapter, we explain how to edit them. There are many more options than we can cover in this one short chapter, but we get you off to a good start. The menus that we explore have many additional options, so explore the available options that you find there.

Working with TableLooks

In this section, we create a similar table three times. The goal is to focus on the editing, not on the table creation, but the table itself is pretty standard.

1. **Open the cars.sav file.**

   The file is not in the SPSS installation directory. You have to download it from this book's companion website.

2. **Choose Analyze ▷ Descriptive Statistics ▷ Crosstabs.**

3. **Choose Number of Cylinders as the Row and Country of Origin as the Column.**

   After selection, your menu should look like Figure 19-1.

Figure 19-1: The Crosstabs dialog box after selection.

	Crosstabs	
car_id	**Row(s):**	Exact...
Miles per Gallon [mpg]	Number of Cylinders [number_of_cyli...	Statistics...
Engine Displacement (cube inches) [...		Cells...
Horsepower [horsepower]	**Column(s):**	Format...
Vehicle Weight (lbs.) [weight]	Country of Origin [country_of_origin]	Style...
Time to Accelerate from 0 to 60 mph...		Bootstrap...
Model Year (modulo 100) [year]	**Layer 1 of 1**	
American car [american_car]	Previous Next	

Display layer variables in table layers

☐ Display clustered bar charts
☐ Suppress tables

Help   Reset   Paste   Cancel   OK

4. Click the Cells button.

The Crosstabs: Cell Display dialog box appears.

5. In the Percentages section, select the Row check box; in the Counts section, deselect the Observed check box (see Figure 19-2).

6. Click Continue.

7. Click OK.

The crosstabulation table between Number of Cylinders and Country of Origin is created.

8. Right-click the crosstabulation table between Number of Cylinders and Country of Origin and choose Edit Content ⇨ In Separate Window, as shown in Figure 19-3.

You're are in editing mode and ready to go.

You can also enter editing mode by double-clicking. By default, however, this won't put you in a new window. The same editing features are available, but editing is done in the main output window.

9. Choose Format ⇨ TableLooks.

The TableLooks dialog box, shown in Figure 19-4, appears.

Make sure you're looking in the menus for the new window that you've opened. Data windows, output windows, Syntax windows, and editing windows all have somewhat different menu choices.

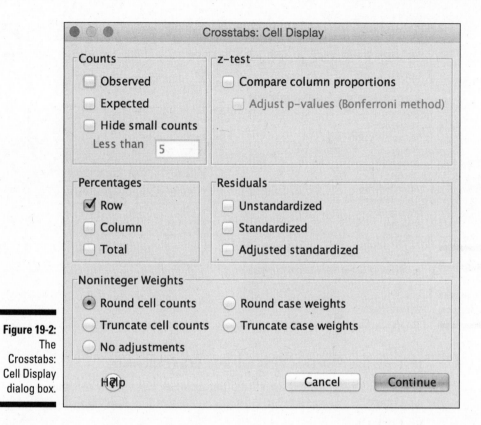

Figure 19-2:
The
Crosstabs:
Cell Display
dialog box.

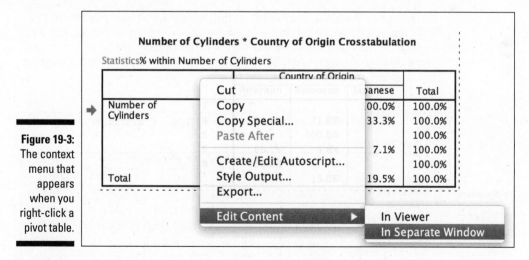

Figure 19-3:
The context
menu that
appears
when you
right-click a
pivot table.

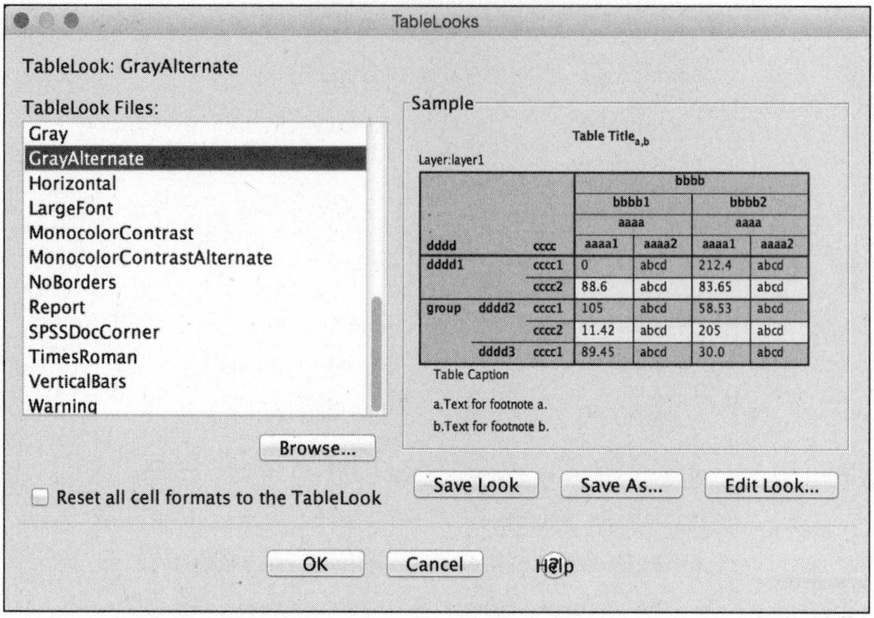

Figure 19-4:
The
TableLooks
dialog box.

10. **From the TableLook Files list, select GrayAlternate.**

    You can play around with the other choices, but Figure 19-5 shows this choice.

11. **Click OK.**

    The result should appear in the output window and look like Figure 19-5. When you close the editing window, you do not lose the output. Rather, it stays in the output window like all the other output. You could try to get away with editing more than one piece of output at a time, but it uses up computer resources so it is best to edit just one.

Number of Cylinders * Country of Origin Crosstabulation

% within Number of Cylinders

		Country of Origin			
		American	European	Japanese	Total
Number of Cylinders	3			100.0%	100.0%
	4	34.8%	31.9%	33.3%	100.0%
	5		100.0%		100.0%
	6	88.1%	4.8%	7.1%	100.0%
	8	100.0%			100.0%
Total		62.5%	18.0%	19.5%	100.0%

Figure 19-5:
A crosstab table with a new appearance.

Style Output

The Style Output feature is relatively new, and quite powerful. This example shows how to highlight certain cells in your output and leave the rest unchanged.

TIP

Menus in SPSS are "sticky," meaning they stay the same unless you reset. So, if you're working through this section immediately after the preceding section, most of these steps are already done.

1. **Open the cars.sav file.**

   The file is not in the SPSS installation directory. You have to download it from this book's companion website.

2. **Choose Analyze ⇨ Descriptive Statistics ⇨ Crosstabs.**

3. **Choose Number of Cylinders as the Row and Country of Origin as the Column.**

   After selection, your menu should look like Figure 19-1.

4. **Click the Cells button.**

   The Crosstabs: Cell Display dialog box appears.

5. **In the Percentages section, select the Row check box; in the Counts section, deselect the Observed check box (refer to Figure 19-2).**

6. **Click Continue.**

7. **Click OK.**

   The crosstabulation table between Number of Cylinders and Country of Origin is created.

8. **Right-click the resulting output and choose Style Output, as shown in Figure 19-6.**

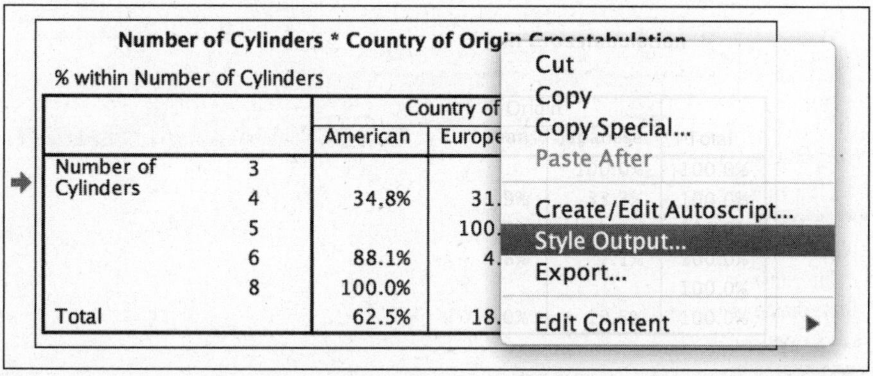

Figure 19-6:
The Style Output context menu option.

Number of Cylinders * Country of Origin Crosstabulation

% within Number of Cylinders

		Country of		
		American	Europe	
Number of Cylinders	3			
	4	34.8%	31.	
	5		100.	
	6	88.1%	4.	
	8	100.0%		
Total		62.5%	18.	

Cut
Copy
Copy Special...
Paste After

Create/Edit Autoscript...
Style Output...
Export...

Edit Content ▶

9. **When the initial dialog box appears, click Continue.**

10. **Scroll down to the Conditional Styling row and click the small square button with three dots.**

    The dialog box is shown in Figure 19-7.

11. **In the Table Style dialog box (shown in Figure 19-8), Choose Percent in the Value column, and then click OK.**

    There are a lot of interesting options here, but the defaults are pretty good. Because our chart is all percentages, we've gone with a percent. The default is to highlight above 50%, and that works well for us.

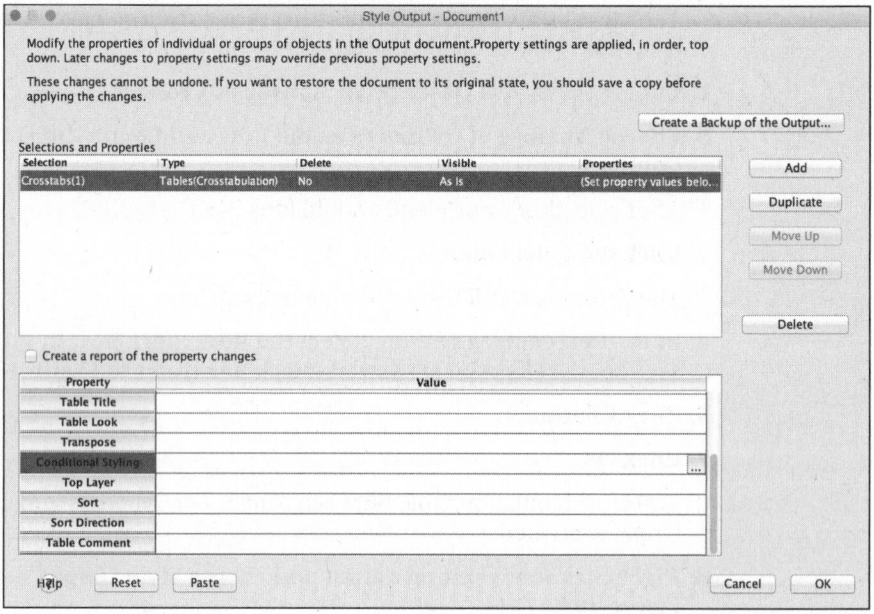

Figure 19-7:
The Style Output main dialog box.

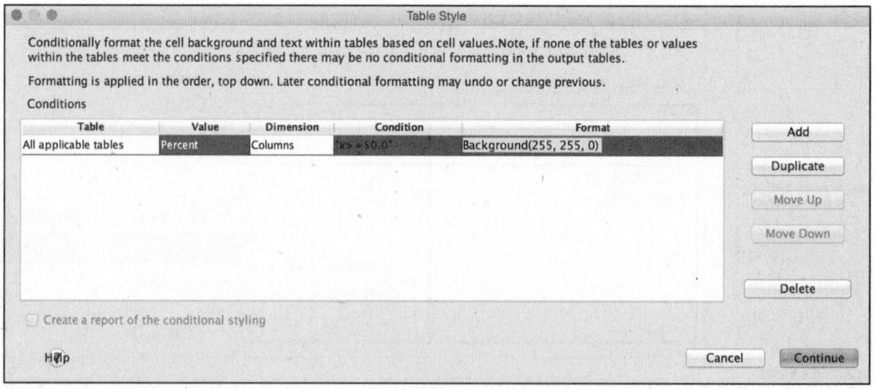

Figure 19-8:
The Table Style dialog box.

12. **Click Continue.**

13. **Click OK.**

The chart has been modified with the percent above 50% highlighted (see Figure 19-9).

number_of_cylinders Number of Cylinders * country_of_origin Country of Origin Crosstabulation

% within number_of_cylinders Number of Cylinders

| | | country_of_origin Country of Origin | | | |
		American	European	Japanese	Total
number_of_cylinders Number of Cylinders	3			100.0%	100.0%
	4	34.8%	31.9%	33.3%	100.0%
	5		100.0%		100.0%
	6	88.1%	4.8%	7.1%	100.0%
	8	100.0%			100.0%
Total		62.5%	18.0%	19.5%	100.0%

Figure 19-9: Table with highlighted percentages.

Pivoting Trays

This is where it all started. This was one of the first benefits of SPSS pivot tables since they were first added to the interface way back in 1995. Pivot tables allow you to rearrange your output in all kinds of ways. Get the hang of this, and you may find yourself taking advantage of it quite often.

In this section, we start with the same table in this example, but we make a major modification to the structure this time.

1. **Open the `cars.sav` file.**

   The file is not in the SPSS installation directory. You have to download it from this book's companion website.

2. **Choose Analyze ⇨ Descriptive Statistics ⇨ Crosstabs.**

3. **Choose Number of Cylinders as the Row and Country of Origin as the Column.**

   After selection, your menu should look like Figure 19-1.

4. **Click the Cells button.**

   The Crosstabs: Cell Display dialog box appears.

5. **In the Percentages section, select the Row and Column check boxes; in the Counts section, select the Observed check box.**

6. **Click Continue.**

7. **Click OK.**

   The result is a larger and more detailed table as shown in Figure 19-10.

8. **Right-click the pivot table and choose Edit Content ⇨ In a Separate Window menu.**

9. **Choose Pivot ⇨ Pivoting Trays, as shown in Figure 19-11.**

   The option appears as a Pivoting Trays button in the Windows menu.

Number of Cylinders * Country of Origin Crosstabulation

			Country of Origin			Total
			American	European	Japanese	
Number of Cylinders	3	Count	0	0	4	4
		% within Number of Cylinders	0.0%	0.0%	100.0%	100.0%
		% within Country of Origin	0.0%	0.0%	5.1%	1.0%
	4	Count	72	66	69	207
		% within Number of Cylinders	34.8%	31.9%	33.3%	100.0%
		% within Country of Origin	28.5%	90.4%	87.3%	51.1%
	5	Count	0	3	0	3
		% within Number of Cylinders	0.0%	100.0%	0.0%	100.0%
		% within Country of Origin	0.0%	4.1%	0.0%	0.7%
	6	Count	74	4	6	84
		% within Number of Cylinders	88.1%	4.8%	7.1%	100.0%
		% within Country of Origin	29.2%	5.5%	7.6%	20.7%
	8	Count	107	0	0	107
		% within Number of Cylinders	100.0%	0.0%	0.0%	100.0%
		% within Country of Origin	42.3%	0.0%	0.0%	26.4%
Total		Count	253	73	79	405
		% within Number of Cylinders	62.5%	18.0%	19.5%	100.0%
		% within Country of Origin	100.0%	100.0%	100.0%	100.0%

Figure 19-10: A cross-tab table with three statistics.

Figure 19-11:
The Pivot
menu in
the editing
window.

Pivot	Format	Analyze	Graph
Reorder Categories ▶			
Transpose Rows and Columns			
Pivoting Trays			
Go to Layers...			

10. **Click the Statistics label in the Pivoting Tray and drag it to a new position in the Layer area.**

    The "before" version is shown in Figure 19-12, and the "after" version is shown in Figure 19-13.

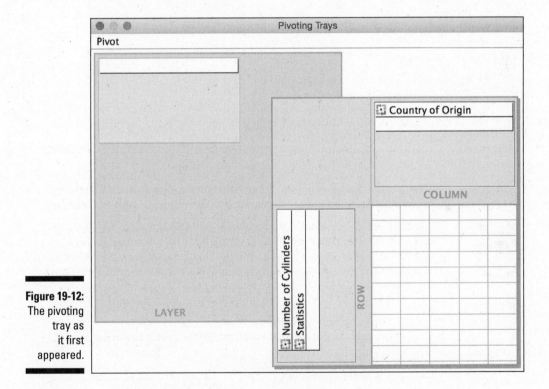

Figure 19-12:
The pivoting
tray as
it first
appeared.

11. **Close the Pivoting Tray window.**

12. **Click the new statistics menu directly on the pivot table, as shown in 19-14.**

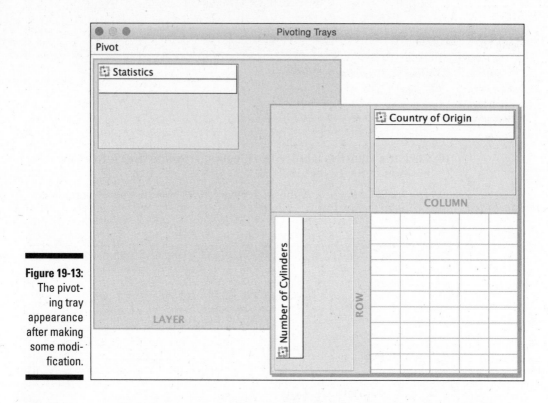

The whole idea of a layer is to have three versions stacked like three sheets of paper. You can choose whichever one you want to display.

One of the Table format options allows you to print all layers, so you actually can use these features to easily have lots of related tables printed to paper or to a PDF. If you like this feature, you don't have to worry about your boss, or coworker, or your classmate not being able to choose the three different versions.

Number of Cylinders * Country of Origin Crosstabulation

Statistic... ✓ Count
% within Number of Cylinders
% within Country of Origin

		American	European	Japanese	Total
Number of Cylinders	3	0	0	4	4
	4	72	66	69	207
	5	0	3	0	3
	6	74	4	6	84
	8	107	0	0	107
Total		253	73	79	405

Figure 19-14:
Choosing the layer using the new menu.

Part VII

Programming SPSS with Command Syntax

See how to work with Graphics Production Language in an article at
www.dummies.com/extras/spss.

In this part . . .

- Get a better sense of what Syntax is all about.
- Try out pasting Syntax, the ultimate Syntax shortcut.
- Go a tiny bit beyond the basics and learn how to write your own Syntax.
- Decide whether learning Syntax is a good choice for you.

Chapter 20

Getting Acquainted with Syntax

*A*lmost everything that happens in SPSS is the result of Command Syntax running behind the scenes. Whenever you use the menu to specify a set of options and then click OK, Command Syntax is generated and executed. You've seen examples of the Command Syntax language — Syntax, for short — appearing at the top of the SPSS Statistics Viewer window every time a command runs. The Help menu has the Command Syntax Reference, which is a real door stopper at more than 2,000 pages. Don't try reading that until you read this chapter and the next one. This chapter describes some language fundamentals and explains why it's helpful. The next chapter gets into more of the details.

So, why use Syntax? First, it's a lot easier than you might think, but no tip or trick in SPSS is worth it unless it improves your results or makes your life easier. Syntax may just be able to do both. Second, there are four main application areas where Syntax might be a better option than menus:

✔ Pasting

✔ Labeling

✔ Repetitive output

✔ Opening and saving files

All four are very easy, and we explain all four in this chapter using a very simple web survey as a case study. Many folks check for survey results from these kinds of surveys every day when new results come in. We show you how to deal with a daily routine like that more efficiently.

The last area — opening and saving files — may surprise you, at first, because it's so easy to do with the menus. You wouldn't switch to Syntax just for that one, but if you do work in Syntax, you can and should open and save your files in Syntax when you do work in Syntax. Otherwise, it's potentially the only manual step in an automated process.

Pasting

Here's the secret to learning Syntax, but this is a secret you can tell all your friends: *You usually don't have to type anything.* This also means you usually don't have to look up commands in a book.

How can that be? It all looks so complicated at first. The secret is the Paste button. There is a button in almost every menu that writes the Syntax for you. You just have to save it for the next time you need it.

A super-easy example is a calculation like the ones we show you in Chapter 9. One of those functions is the LENGTH function; we wrote the formula you see in Figure 20-1.

Now, use that Paste button in the bottom left of Figure 20-1, and — *voilà!* — you get the following result in the Syntax editor window:

```
COMPUTE Length=length(comment).
EXECUTE.
```

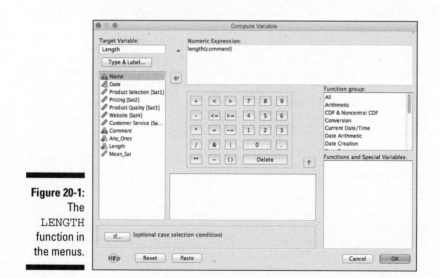

Figure 20-1:
The
LENGTH
function in
the menus.

Whenever you want to run it, you can simply choose Run⇨All from the Syntax editor menu, or you can highlight just a portion and choose Data⇨Select. It's really that simple.

Now, there is more to Syntax than that. We have an entire second chapter on it (and there are entire books dedicated to it), but as far as pasting goes, it's pretty simple.

Make sure you don't confuse the Paste button with Ctrl+V–type operations in SPSS Statistics or Microsoft Excel. The Paste button isn't the same as Copy and Paste. It generates Syntax code for you, and you need to use the Paste button as shown in Figure 20-1.

The EXECUTE looks so official that you'd probably guess that you need it. The truth is, you usually don't. In this case, we need it because nothing generates output after EXECUTE. If you had any kind of table- or graph-generating command after it — and you almost always do — you could (and should) delete it. More on this in the next chapter.

Labeling

This is one of the very easiest Syntax commands to learn how to type directly. Open the Syntax window, and try the following commands:

```
VARIABLE LABELS
Sat1 "Product Selection"
Sat2 "Pricing"
Sat3 "Product Quality"
Sat4 "Website"
Sat5 "Customer Service".
VALUE LABELS Sat1 TO Sat5
1 'Very Dissatisfied'
2 'Dissatisfied'
3 'Neither Satisfied nor Dissatisfied'
4 'Satisfied'
5 'Very Satisfied'.
```

It's really difficult to argue that this is any harder than doing it in the menus, especially because the label phrases are often lying about in a text editor somewhere already. This way, you can rerun it any time you need to. The menus are great, but this is pretty easy, and it may turn out to be considerably faster for you if you use this data repeatedly. Your coworkers will love it because they can borrow it.

Repeatedly Generating the Same Reports

Imagine that every morning your boss wants a quick rundown on the web survey activity that's taken place since the previous report. She's busy, so she wants it organized in a very particular way:

- ✔ She wants to know which aspects of the business folks are the most happy with at the top.

- ✔ She wants to read comments left by survey takers, but she doesn't want blank comments in the way.

- ✔ She wants the comments from the happy folks at the top.

So, the boss wants the DESCRIPTIVES report to look like Figure 20-2 and the simple little report to look like Figure 20-3.

Figure 20-2:
The
Satisfaction
variables
reported in
order of
satisfaction.

Descriptives

Descriptive Statistics

	N	Minimum	Maximum	Mean
Product Selection	10	1	5	4.50
Customer Service	9	1	5	4.44
Product Quality	9	1	5	4.22
Pricing	10	1	5	4.20
Website	9	1	5	4.11
Valid N (listwise)	9			

Figure 20-3:
A Case
Summaries
showing
only cases
with com-
ments and
broken
down by
Any_
Ones.

Case Summaries[a]

			Name	Comment		Mean_Sat
Any_Ones	.00	1	Ann	None		5.00
		2	Erin	Your company is great!		5.00
		3	Frank	I did not receive my order yet.		5.00
		4	Irisa	I've noticed a decline since last year.		4.60
		5	Bob	Not at this time		4.40
		Total N	5		5	5
	1.00	1	Cara	Good products, but confusing website.		4.00
		2	Hank	You guys have a great product.		1.00
		Total N	2		2	2
	Total	N	7		7	7

a. Limited to first 100 cases.

The following code would do all that every morning, and it wouldn't require reproducing the steps. Just choose Run ⇨ All from the Syntax window. Easy.

This code works on the Web Survey2.sav dataset, and the code has been saved as Web Survey.sps.

```
DESCRIPTIVES VARIABLES=Sat1 Sat2 Sat3 Sat4 Sat5
/STATISTICS=MEAN MIN MAX
/SORT=MEAN (D).
SORT CASES BY Mean_Sat(D).
TEMPORARY.
SELECT IF length >0.
SUMMARIZE
/TABLES=Name Comment Mean_Sat BY Any_Ones
/FORMAT=VALIDLIST NOCASENUM TOTAL LIMIT=100
/TITLE='Case Summaries'
/MISSING=VARIABLE
/CELLS=COUNT.
```

At first glance, it may look like this would take a substantial amount of knowledge to put together, but you may be surprised. The DESCRIPTIVES command can be easily pasted just like we saw at the opening of the chapter. Same thing with SORT and SUMMARIZE. The SUMMARIZE command is found under Analyze⇨Reports⇨Case Summaries. The other commands, TEMPORARY AND SELECT, we just typed, and we tell you more about both of them and typing commands in the Syntax window, too, in the next chapter. The main point for now is that SPSS will do 80% of the typing for you. Just use the Paste button to save your work. After you've tested it, and you know that it works, save the whole thing for the next time.

Opening and Saving Files

You can write data to files and read data from files. Output is performed by using either SAVE or EXPORT. The keyword IMPORT can be used to read data from a file, but the simplest way to read files is to read SPSS-formatted files using GET. If you're doing your analysis every day, you may as well open your file automatically. There is no need to do that step manually.

Just like virtually all the windows, the GET and SAVE commands allow you to paste. You shouldn't have to type these commands or the file paths very often if you prefer the menus.

GET

Whenever you choose File⇨Open ⇨ Data, SPSS issues a GET command to open an SPSS-formatted file and load it into SPSS. If you've loaded a file using the menu this way, you've noticed in SPSS Statistics Viewer the GET command that loads the file. For example, the following program opens and loads the file named Cars.sav:

```
GET
FILE='C:\Program Files\SPSS\Cars.sav'.
DATASET NAME DataSet2 WINDOW=FRONT.
```

This command loads the data from the file, names it DataSet2, and opens a new SPSS main window to display the data from the file — in front of all the other windows. You never have to load a file with the menu — you can load any file from within a Syntax program by specifying its name as the first argument to a GET command.

The quotes around the filename are optional unless a blank is embedded in the name.

You don't have to load the entire contents of the file. If you want to omit certain variables, you can name them as part of the command, as in the following:

```
GET FILE='C:\Program Files\SPSS\Cars.sav' /DROP=MPG engine_size.
```

You can even change the names of some variables. For example, the following changes MPG to MILESPERGALLON:

```
GET FILE='C:\Program Files\SPSS\Cars.sav' /RENAME=MPG=MILESPERGALLON.
```

SAVE

The SAVE command has the same result as choosing File⇨Save As and entering a filename. It writes the data to a file in the standard SPSS format. An example of the command follows:

```
SAVE OUTFILE='C:\Program Files\SPSS\Cars.sav'.
```

You have some options. You can specify DROP and RENAME the same as you can with the GET command. You can also compress the output file with the following option:

```
/COMPRESSED
```

Chapter 21

Adding Syntax to Your Toolkit

*I*n the last chapter, you see that Syntax can save you time (avoiding repetitive steps and making it easier for others to know exactly what you've done) and that ultimately it's really pretty easy to get started. When you incorporate Syntax into your routine, you'll want to start expanding your use of it — as long as it continues to save you time and effort.

Some Syntax commands can't be pasted from the menus using the Paste button. Some are just easier to type. If you're going to start typing some Syntax, you'll have to learn a little bit more about the grammar and the rules of Syntax.

You can always choose to do most things in SPSS with the menus. It's okay to use the menus for a while before you master Syntax. There is an old saying, "When you're ready, your teacher will appear." So, when you're ready, Syntax — and this chapter — will be waiting for you.

Your Wish Is My Command

A single Syntax language instruction can be very simple, or it can be complex enough to serve as an entire program. A single instruction consists of a command followed by arguments to modify or expand the actions of the command. For example, the following Syntax command generates a report:

```
REPORT /FORMAT=LIST /VARIABLES=MPG.
```

The first thing you probably noticed is that the command is written in all uppercase. That's tradition — not a requirement. You can write in lowercase (or even mixed case) if you want. The old-school way of writing it, dating back to when everything was typed, is to write commands in uppercase, and variable names in lowercase. The Syntax window's autocomplete function will write commands in uppercase. Notice, too, that the end of the list of arguments is terminated by a single period; the terminator must be there or else SPSS will complain.

Now, about those forward slashes and equal signs: Sometimes you need them, and sometimes they're optional. Always use them and you won't have any trouble. The presence of slashes and equal signs reduces ambiguity for both you and SPSS. Also, commands can be abbreviated as long as you have at least three letters that uniquely identify the command. Abbreviating commands was a popular strategy when everything was typed, but we can't think of a single reason to abbreviate anything, especially with the autocomplete feature. Figuring out how to abbreviate a command is more work than just typing it, and abbreviation makes the program harder to figure out later.

The command in this example is REPORT, which causes text to be written to SPSS Statistics Viewer. In fact, all output produced by running Syntax programs goes to SPSS Statistics Viewer. The FORMAT specification tells REPORT to make a list of the values. The VARIABLES specification tells REPORT which variables are to be included in the list.

Commands can begin anywhere on a line and continue for as many lines as necessary. That's why SPSS is so persnickety about that terminator (the period) — it's the only way it has of detecting the end of a command. If you forget it, SPSS may think that the additional lines belong to an earlier line. If the syntax turns bright red, it's a bad sign. Very bad. Try deleting a period, and the colors will change. Trouble. Big trouble. What you're hoping for is for the command to turn blue. Subcommands will be green, and keywords will be maroon. All the stuff that is unique to your program, like your dataset's variables, will be a plain old black text. Table 21-1 lists all the command types and the colors in the Syntax Editor.

Table 21-1	Color Coding in the Syntax Editor
Syntax Command Type	*Color in the Syntax Editor*
Command	Blue
Subcommand	Green
Keyword	Maroon
Other (including variable names)	Black
Error	Red

Using Keywords

All the commands in Syntax are keywords in the language. A *keyword* is a word already known to the language; it has a predefined action. The variable names you define are not keywords, but SPSS can tell which is which by the way you use them. That is, you can name one of your variables the same name as one of the keywords, and SPSS can tell what you mean by how you use the word. Usually.

The names of commands, subcommands, and functions are all keywords — and there are lots of them — but they aren't reserved and you can use them freely. For example, you could have variables named `format` and `report`, and you could use the following Syntax command to display a list of their values:

```
REPORT /FORMAT=LIST /VARIABLES=REPORT FORMAT.
```

Don't try to name variables AND, OR, or NOT. These logical operators are keywords in the Syntax language and are *also* reserved words. If you try to use a reserved word as a variable name, SPSS will catch it and tell you that you can't do it. Relational operators are used in the Syntax language to compare values and are also reserved words. The relational operators are EQ, NE, LT, GT, LE, and GE. Some other reserved words are ALL, BY, TO, and WITH. (These operators are discussed in more detail later in this chapter.)

Working with Variables and Constants

Most of the values used in Syntax come from the variables in the dataset you currently have loaded and displayed in SPSS. You simply use your variable names in your program, and SPSS knows where to go and get the values. Some other variables are already defined, and you can use them anywhere in a program. Predefined variables, which are called *system variables,* all begin with a dollar sign ($) and already contain values. The system variables are listed in Table 21-2.

Table 21-2	System Variables
Variable Name	*Description*
$CASENUM	The current case number. It's the count of cases from the beginning case to the current case.
$DATE	The current date in international date format with a two-digit year.

(continued)

Table 21-2 *(continued)*

Variable Name	Description
$DATE11	The current date in international date format with a four-digit year.
$JDATE	The count of the number of days since October 14, 1582 (the first day of the Gregorian calendar).
$LENGTH	The current page length.
$SYSMIS	The system missing value. This prints as a period or whatever is defined as the decimal point.
$TIME	The number of seconds since midnight October 14, 1582 (the first day of the Gregorian calendar).
$WIDTH	The current page width.

When a Syntax program executes, it's associated with the currently loaded dataset and uses its variable names and values. This can get confusing when you have more than one dataset open. If SPSS claims that there is no variable with that name, make sure that the correct dataset is active.

Declaring Data

You can define variables and their values right in the Syntax window. You might wonder why you would do this. Why not just have a data file? This is a great way to ask for advice and to prototype calculations. You can send the DATA LIST command with just a few rows of data when you ask a colleague for help. Just copy and paste it right into an email along with your question and the code that you're trying to fix.

To do so, you create a DATA LIST, which defines the variable names, and follow it with the list of values between BEGIN DATA and END DATA commands. The following example creates three variables (ID, SEX, and AGE) and fills them with four instances of data:

```
DATA LIST / ID 1-3 SEX 5 (A) AGE 7-8.
BEGIN DATA.
001 m 28
002 f 29
003 f 41
004 m 32
END DATA.
PRINT / ID SEX AGE.
EXECUTE.
```

The DATA LIST command defines the variables. The first variable is ID. Its values are found in the input stream in columns 1 through 3; therefore it's defined as being three digits long. It has no type definition, so it defaults to numeric. The second variable is named SEX. It is one character long, and its values are in column 5 of the input. Its type is declared as alpha (A), so it's declared as a one-character string. The third variable, AGE, is two digits long, is a numeric value, and has its values in columns 7 and 8 of the input.

The BEGIN DATA command comes immediately after the DATA LIST command and marks the beginning of the lines of data — each line is a case. If you've ever wondered what it was like to place data on punched cards, this is it. The fundamental design of SPSS is that old. This form of data entry still works, but this is the old way of getting data into a program. When this list of commands is executed, a normal SPSS window appears, showing a dataset with the variable names and values.

You can do all your processing this way, if you prefer. But you don't have to do it by column numbers. You can enter the data in a comma-separated list, as follows:

```
DATA LIST LIST (",")/ ID SEX AGE.
BEGIN DATA.
1,1,28
2,2,29
3,2,41
4,1,32
END DATA.
PRINT / ID SEX AGE.
EXECUTE.
```

END DATA must begin in the first column of a command line. It's the only command in Syntax that has this requirement.

Commenting Your Way to Clarity

You can insert descriptive text, called a *comment,* into your program. This text doesn't do anything except help clarify how the program works when you read (or somebody else reads) your code. You start a comment the same way you start any other command: on its own line, using the keyword COMMENT or an asterisk. The comment is terminated by a period. Here's an example:

```
COMMENT This is a comment and will not be executed.
```

An asterisk can be used with the same result, which is the way that everyone really does it:

```
* This is a comment placed here for the purpose of
describing what is going on, and it continues until
it is terminated.
```

You can also put comments on the same line as a command by surrounding them with /* and */. A comment like this can be inserted anywhere inside the command where you'd normally put a blank. For example, you could put a comment at the end of a command line like this:

```
REPORT /FORMAT=LIST /VARIABLES=AGE /* The comment */.
```

It is important to note that the command is terminated with a period, but the period comes after the comment because the comment is part of the statement. If you forget, the next line will get swallowed up into the comment and ignored. The following line will not be color-coded correctly either, which may help you catch your mistake. Watch out.

Executing Commands

Commands are executed one at a time, starting from the top of the program. The order is important. In particular, if a variable has not been created yet, you can't use it. For the most part, the order is intuitive; you don't have to think much about what exists and what doesn't.

Some statements don't execute right away. Instead, they're stored for later execution. This makes SPSS run faster, so it's all for a good reason, but most folks don't know how it works. This is normally of no consequence because the statements will be executed when their result is needed. But you should be aware this is going on because it can cause surprises in some circumstances. For example, the PRINT command has a delayed execution:

```
PRINT / ALL.
```

This is a command to print the complete list of values for every case in your dataset. It will print all the values, or by naming variables it can be instructed to print values of only the ones you choose. However, the PRINT command doesn't do it right away. It stores the instruction for later. Commands like this are called *transformations*. As you might guess, all the commands in the Transform menu are of this type. Commands like COMPUTE, COUNT, and RECODE are transformations.

When your program comes to a command that executes immediately, the stored commands are executed first. That works fine as long as there's another statement to be executed, but if the PRINT statement is the last one in your program, nothing happens. That is, nothing happens until you run another program, and then the stored statement becomes the first one executed.

But there is an easy fix that you may see in some Syntax programs. All you need to do is end your program this way:

```
PRINT / ALL.
EXECUTE.
```

All the EXECUTE command does is execute any statements that have been stored for future execution. You'll see programs written by others who do it this way, but you generally don't want to solve the problem this way. *Procedure commands* (commands that generate output) will accomplish the same thing. So, just put any old procedure that you had to do anyway after your transformation, like FREQUENCIES, as shown below:

```
PRINT / ALL.
FREQUENCIES ALL.
```

For the PRINT command, there is another option. The LIST command does the same thing the PRINT command does, but it executes immediately instead of waiting until the next command:

```
LIST / ALL.
```

A number of commands have a Transform version and a Procedure version. For instance, SAVE is a Procedure, and XSAVE is a Transformation. (The Command Syntax Reference, which is located under the Help menu, has lists of which commands are which.) This execution delay may seem odd at first, but there's a really good reason for it: If SPSS executed every line one at a time, it would have to reread the data for every line and would be very slow. A little tricky, but very important information.

Remember that the transformations are delayed, but the procedures happen right away. Your Syntax programs will run best (and fastest) when you try to put all your transformations at the top, and all your procedures at the bottom. Transformations Pending is an error that you'll get when you've managed to end your program with a transformation.

Controlling Flow and Executing Conditionals

Unless you specify otherwise, a program starts at the top and executes one statement at a time through your program until it reaches the bottom, where it stops. But you can change that. Situations come up where you need to execute a few statements repeatedly, or maybe you want to conditionally skip one or more statements. In either case, you want program execution to jump from one place to another under your control. What you're really trying to do is say that certain cases will be treated one way (by certain lines) and other cases will be treated another way (by other lines).

IF

You use the IF command when you have a single statement you want to execute only if conditions are right. For example:

```
IF (AGE > 20) GROUP=2.
```

This statement asks the simple question of whether AGE is greater than 20. If so, the value of GROUP is set to 2. We could've used the GT keyword in place of the > symbol. Table 21-3 lists the relational operators you can use to compare numbers.

Table 21-3	Relational Operators	
Symbol	Alpha	Definition
=	EQ	Is equal to
<	LT	Is less than
>	GT	Is greater than
<>	NE	Is not equal to
<=	LE	Is less than or equal to
>=	GE	Is greater than or equal to

You can also combine the relational expressions with logical operators to ask longer and more complex questions. Here's an example:

```
IF (AGE > 20 AND SEX = 1) GROUP=2.
```

This statement asks whether AGE is greater than 20 and SEX is equal to 1. If so, GROUP is set to 2. The logical operators are listed in Table 21-4.

Table 21-4	Logical Operators	
Symbol	*Alpha*	*Definition*
&	AND	Both relational operators must be true.
|	OR	Either relational operator can be true.
~	NOT	Reverses the result of a relational operator.

You should use parentheses to organize expressions so there is no ambiguity about what is being compared. When you construct a complicated conditional expression, it's easy to lose track of your original line of scrimmage.

You have to write your expressions so the computer knows what you're talking about. Spell them out. For example, IF (A LT B OR GT 5) is not valid. It can be written IF ((A LT B) OR (A GT 5)), which is a longer form but has a clearer meaning.

You can compare strings to strings and numbers to numbers, but you can't compare strings to numbers.

DO IF

The DO IF statement works the same way as the IF statement, but with DO IF you can execute several statements instead of just one. Because you can enter several statements before the terminating END IF, the END IF is required to tell SPSS when the DO IF is over. The following is an example with two statements:

```
DO IF (AGE < 5).
COMPUTE YOUNG = 1.
COMPUTE SCHOOL = 0.
END IF.
```

In addition to having the option of including a number of statements at once, you can use DO IF to test several conditions in a series — and execute only the statements of the first true condition(s) by using ELSE IF:

```
DO IF (AGE < 5).
COMPUTE YOUNG = 1.
ELSE IF (AGE < 9).
COMPUTE YOUNG = 2.
ELSE IF (AGE < 12).
COMPUTE YOUNG = 3.
END IF.
```

SELECT IF

The SELECT IF statement is not really flow control, but it works that way. You can use it to remove specific cases, and, as a result, include only the cases you want in your analysis. For example, the following sequence of commands prints only the age values greater than 40:

```
SELECT IF (AGE > 40).
PRINT / AGE.
EXECUTE.
```

Watch out, though. If you save your dataset right after this command, you'll lose data! Any of the logical operators and relational operators that can be used in other IF statements can be used in SELECT IF statements. A really powerful and popular way to modify this is the TEMPORARY command.

```
TEMPORARY.
SELECT IF (AGE > 40).
PRINT / AGE.
EXECUTE.
```

The SELECT IF will work only until it hits the EXECUTE (or any other procedure). Then SPSS immediately goes back to using all the data. Much better, because it's less risky. Less risky is nice.

If you have any procedures that you have to do anyway, you can (and should) delete the EXECUTE command. Just make sure that some procedure — any procedure — comes after the transformation. Pretty much any command that generates output (table or graph) is a procedure.

Part VIII
The Part of Tens

Find out about the top ten new SPSS features in a free article at www.dummies.com/extras/spss.

In this part . . .

- ✔ Figure out what all those modules are about and if any are right for you.
- ✔ Navigate through all that SPSS stuff on the Internet and get directions to the very best of it.
- ✔ Explore ten ways you can sharpen your skills with SPSS.

Chapter 22

Ten (Or So) Modules You Can Add to SPSS

In This Chapter

▶ Understanding what modules bring to SPSS

▶ Deciding which modules to add to SPSS

. . . .

*I*BM SPSS Statistics comes in the form of a base system, but you can acquire additional modules to add on to that system. If you've installed a full system, you may already have some of these add-ons. Most are integrated and look like integral parts of the base system. Some may be of no interest to you; others could become indispensable. Note that if you have a trial copy of SPSS, it likely has all the modules, including those that you might lose access to when you acquire your own copy. This chapter introduces you to the modules that can be added to SPSS and what they do; refer to the documentation that comes with each one for a full tutorial.

The Advanced Statistics Module

The following is a list of the statistical techniques that are part of the Advanced Statistics module.

- ✔ General Linear Models (GLM)
- ✔ Generalized Linear Models (GENLIN)
- ✔ Linear Mixed Models
- ✔ Generalized Estimating Equations (GEE) Procedures
- ✔ Generalized Linear Mixed Models (GLMM)
- ✔ Survival Analysis Procedures

Making sense of editions

Since IBM purchased SPSS, the concept of "editions" has caused some confusion. Frankly, we have to look it up to remind ourselves which modules are in which edition. The editions are just bundles. And there are a bunch of modules — considerably more than the ten we cover in this chapter. You can look up the information you need on the IBM website at `www-01.ibm.com/software/analytics/spss/products/statistics/modules`. We also explain what's in each edition right here.

It isn't always obvious that you have a module installed. Basically, if you have access to the advanced features, then you have the module. For instance, Generalized Estimating Equations require the Advanced Statistics module.

Sometimes end-users get the impression that "standard" is the entry-level version of SPSS. But that's actually SPSS Base. SPSS Base has no modules. You can get modules as part of a bundle, or you can buy them separately.

Here's a list of all the modules and in which editions they appear:

✔ **Advanced Statistics:** Standard Edition

✔ **Amos:** Premium Edition

✔ **Bootstrapping:** Premium Edition

✔ **Categories:** Professional Edition

✔ **Complex Samples:** Premium Edition

✔ **Conjoint:** Premium Edition

✔ **Custom Tables:** Standard Edition

✔ **Data Preparation:** Professional Edition

✔ **Decision Trees:** Professional Edition

✔ **Direct Marketing:** Premium Edition

✔ **Exact Tests:** Premium Edition

✔ **Forecasting:** Professional Edition

✔ **Missing Values:** Professional Edition

✔ **Neural Networks:** Premium Edition

✔ **Regression:** Standard Edition

✔ **Sample Power:** Premium Edition

✔ **Visualization Designer:** Premium Edition

Note: Premium includes all the features in Professional and Standard; Professional includes all the features in Standard.

These procedures are among the most advanced in SPSS, but some of them are quite popular. For instance, Hierarchical Linear Modeling (HLM), part of Linear Mixed Models, is popular in education research. HLM models are statistical models where parameters vary at more than one level. For instance, students vary and schools vary, and in an HLM model you have information at both levels.

The key point is that this whole module is about specialized techniques that you need to use if you don't meet the assumptions of plain vanilla regression and analysis of variance (ANOVA). These techniques are more of an ANOVA flavor. Survival Analysis is so-called "time-to-event" modeling, like estimating time to death after diagnosis.

The Custom Tables Module

This has been the most popular module for years, and for good reason. If you need to squeeze a lot of information into a report, you need this module. For instance, if you do survey research and you want to report on the entire survey in tabular form, this module comes to your rescue. Picture your entire dataset summarized in an appendix. It isn't merely a convenience. If you need this kind of summary, get this module.

The Regression Module

The following is a list of the statistical techniques that are part of the Regression module.

- Multinomial and Binary Logistic Regression
- Nonlinear Regression (NLR) and Constrained Nonlinear Regression (CNLR)
- Weighted Least Squares Regression and Two-Stage Least Squares Regression
- Probit Analysis

In some ways, this module is like the Advanced Stats module — you use these techniques when you don't meet the standard assumptions — except with this module, the techniques are fancy variants of regression when you can't do Ordinary Least Squares Regression. Binary Logistic Regression is very popular and used quite often. It's used when your dependent variable has two categories — for example, stay or go (churn), buy or not buy, or get a disease or not get a disease.

The Categories Module

The Categories module is designed to enable you to reveal relationships among your categorical data. To help you understand your data, the Categories module uses perceptual mapping, optimal scaling, preference scaling, and dimension reduction. Using these techniques, you can visually interpret the relationships among your rows and columns.

Categories performs its analysis and displays results so you can understand ordinal and nominal data. It uses procedures similar to conventional regression, principal components, and canonical correlation. It performs regression using nominal or ordinal categorical predictor or outcome variables.

The procedures of the Categories module make it possible to perform statistical operations on categorical data:

- ✔ Using the scaling procedures, you can assign units of measurement and zero-points to your categorical data, which gives you access to new groups of statistical functions because you can analyze variables using mixed measurement levels.

- ✔ Using correspondence analysis, you can numerically evaluate similarities among nominal variables and summarize your data according to components you select.

- ✔ Using nonlinear canonical correlation analysis, you can collect variables of different measurement levels into sets of their own, and then analyze the sets.

You can use this module to produce a couple of very useful tools:

- ✔ **Perceptual maps:** High-resolution summary charts that serve as graphic displays of similar variables or categories. They give you insights into relationships among more than two categorical variables.

- ✔ **Biplots:** Summary charts that make it possible to look at the relationships among products, customers, and demographic characteristics.

The Data Preparation Module

Let's face it: Data preparation is no fun. We'll take all the help we can get with it. No module will eliminate all the work for the human in this human–computer partnership, but the Data Preparation module is designed to eliminate some of the routine, predictable aspects. It helps you process your rows and columns of data. For your rows of data, it helps you identify outliers that might distort your data. As for your variables, it helps you identify the best ones, and lets you know that you could improve some by transforming them. It also allows you to create special validation rules to speed up your data checks and avoid a lot of manual work. Finally, it helps you identify patterns in your missing data.

The Decision Trees Module

Decision trees are, by far, the most popular and well known of the data mining techniques. In fact, there are entire software products dedicated to this approach. If you aren't sure if you need to do data mining, but you want to try it out, this would be just about the best way because you already know your way around SPSS Statistics. The Decision Trees module doesn't quite have all the features of the decision trees in SPSS Modeler (which is a whole software package dedicated to data mining), but there is plenty here to give you a good start.

What are decision trees? Well, the whole idea is that you have something that you want to predict (the target variable), and lots of variables that could possibly help you do that, but you don't know which ones are most important. SPSS indicates which variables are most important and how the variables interact, and helps you predict the target variable in the future.

SPSS supports four of the most popular algorithms for doing this: CHAID, Exhaustive CHAID, C&RT, and QUEST.

The Forecasting Module

You can use the Forecasting module to rapidly construct expert time-series forecasts. This module includes statistical algorithms you can use to analyze historical data and predict trends. You can set it up to analyze hundreds of different time series at once instead of running a separate procedure for each one.

The software is designed to handle the special situations that arise in trend analysis. It automatically determines the best-fitting autoregressive integrated moving average (ARIMA) or smoothing model. It automatically tests data for seasonality, intermittency, and missing values. The software detects outliers and prevents them from unduly influencing the results. The graphs generated include confidence intervals and indicate the model's goodness of fit.

As you gain experience at forecasting, the Forecasting module gives you more control over every parameter when you're building your data model. You can use the Expert Modeler in the Forecasting module to recommend starting points or to check calculations you've done by hand.

Version 23 has an exciting new algorithm that is part of this module. It's called Temporal Causal Modeling (TCM). This new algorithm attempts to discover key causal relationships in time series data by including only those inputs that have a causal relationship with the target. This differs from traditional time series modeling where you must explicitly specify the predictors for a target series.

The Missing Values Module

The Data Preparation module seems to have missing values covered, but the two modules are actually quite different. The Data Preparation module is really about finding data errors; its validation rules will tell you that a data point just isn't right. On the other hand, the Missing Values module is focused on when there is no data value at all. It attempts to estimate the missing piece of information using other data that you do have. This process is called *imputation,* also known as replacing with an educated guess. All kinds of researchers, data miners, and statisticians can benefit, but if you're a survey researcher, this is really bound to come in handy.

The Bootstrapping Module

Hang on tight, we're going to get a little technical. *Bootstrapping* is a technique that involves "resampling" with replacement. What that means is that the Bootstrapping module picks a case at random, makes note about it, replaces it, and picks another. In this way, it's possible to pick a case more than once or not at all. The net result is another "version" of your data that is similar, but not identical. If you do this 1,000 times (which is the default), you can do some very powerful things indeed.

The Bootstrapping module allows you to build more stable models using your data by overcoming the effect of outliers and other problems in your data. Traditional statistics assumes that your data has a particular distribution, but this technique avoids that kind of assumption. The result is a more accurate sense of what's going on in the population. It is, in a sense, a simple idea, but because it takes a lot of computer horsepower, it's more popular now than when computers were slower.

Bootstrapping is a popular technique outside of SPSS, as well, so you can find articles on the web about the concept. The Bootstrapping module just lets you apply this powerful concert to your data in SPSS Statistics.

The Complex Samples Module

Sampling is a big part of statistics. A *simple random sample* is what we usually think of as a sample — like picking names out of a hat. The hat is our population, and the scraps of paper we pick belong to our sample. Each slip of paper has an equal chance of being picked. Research is often more complicated than that. The Complex Sample module is about more complicated forms of sampling: two stage, stratified, and so on.

Most often, survey researchers need this module, although many kinds of experimental researcher may benefit from it, too. It helps you design the data collection, and then takes the design into account when calculating your statistics. Nearly all statistics in SPSS are calculated with the assumption that the data is a simple random sample. Your calculations can be distorted when this assumption is not met.

The Conjoint Module

The Conjoint module provides a way for you to determine how each of your product's attributes affect consumer preference. When you combine conjoint analysis with competitive market product research, it's easier to zero in on product characteristics that are important to your customers.

With this research, you can determine which product attributes your customers care about, which ones they care about most, and how you can do useful studies of pricing and brand equity. And you can do all this *before* incurring the expense of bringing new products to market.

The Direct Marketing Module

This module is a little different from the others. It's a bundle of related features in a wizardlike environment. It's designed to be one-stop shopping for marketers. The main features are recency frequency monetary (RFM) analysis, cluster analysis, and profiling:

- **RFM analysis:** RFM analysis reports back to you about how recently, how often, and with how much spend your customers bring your business. Obviously, those customers who are currently active, spend a lot, and spend often are your best customers.

- **Cluster analysis:** Cluster analysis is a way of segmenting your customers into different customer segments. Typically, you use this approach to match different marketing campaigns to different customers. For example, a cruise line may try different covers on the travel catalog going out to customers, with the adventurous types getting Alaska or Norway on the cover, and the umbrella-drink crowd getting pictures of the Caribbean.

✔ **Profiling:** Helps you can see what customer characteristics are associated with specific outcomes and in this way you can calculate the propensity score that a particular customer will respond to a specific campaign. Virtually all these features can be found in other areas of SPSS, but the wizardlike environment of the Direct Marketing module makes it easy for marketing analysts who don't happen to have extensive training in the statistics behind the techniques to be able produce useful results.

The Exact Tests Module

The Exact Tests module makes it possible to be more accurate in your analysis of small datasets and datasets that contain rare occurrences. It gives you the tools you need for analyzing such data conditions with more accuracy than would otherwise be possible.

When only a small sample size is available, you can use the Exact Tests module to analyze that smaller sample and have more confidence in the results. Here, the idea is to perform more analyses in a shorter period of time. This module allows you to conduct different surveys rather than spend time gathering samples to enlarge the base of the surveys you have.

The processes you use, and the forms of the results, are the same as those in the base SPSS system, but the internal algorithms are tuned to work with smaller datasets. The Exact Tests module provides more than 30 tests covering all the nonparametric and categorical tests you normally use for larger datasets. Included are one-sample, two-sample, and K-sample tests with independent or related samples, goodness-of-fit tests, tests of independence, and measures of association.

The Neural Networks Module

A *neural net* is a latticelike network of neuronlike nodes, set up within SPSS to act something like the neurons in a living brain. The connections between these nodes have associated *weights* (degrees of relative effect), which are adjustable. When you adjust the weight of a connection, the network is said to learn.

In the Neural Network module, a training algorithm iteratively adjusts the weights to closely match the actual relationships among the data. The idea is to minimize errors and maximize accurate predictions. The computational neural network has one layer of neurons for input, another for output, with one or more hidden layers between them. The neural network is combined with other statistical procedures to provide clearer insight.

Using the familiar SPSS interface, you can mine your data for relationships. After selecting a procedure, you specify the dependent variables, which may be any combination of scale and categorical types. To prepare for processing, you lay out the neural network architecture, including the computational resources you want to apply. To complete preparation, you choose what to do with the output:

✔ List the results in tables.

✔ Graphically display the results in charts.

✔ Place the results in temporary variables in the dataset.

✔ Export models in XML-formatted files.

Amos

Amos is an interactive interface you can use to build structural equation models. Not a true "module," it's standalone software with its own graphical user interface (GUI). Using the diagrams you create with Amos, you can uncover otherwise-hidden relationships and observe graphically how changes in certain values affect other values. You can create a model on nonnumeric data without having to assign numerical scores to the data. You can analyze censored data without having to make assumptions beyond normality.

Amos provides a more intuitive interface than plain SPSS for a certain family of problems. Amos contains structural modeling software that you control with a drag-and-drop interface. Because the interface is intuitive, you can create models that come closer to the real world than the multivariate statistical methods of SPSS. You set up your variables, and then you can perform analyses using hypothetical relationships.

Amos enables you to build models that more realistically reflect complex relationships with the ability to use observed variables, such as survey data or latent variables like "satisfaction" to predict any other numeric variable. *Structural equation modeling,* helps you gain additional insight into causal models and the strength of variable relationships.

The Sample Power Module

The Sample Power module was developed in conjunction with the late Jacob Cohen. Cohen was a contemporary statistics powerhouse, deservedly famous for his books and articles, and largely responsible for drawing more attention to Type II error. The idea is that our university training emphasizes avoiding Type I error to such a degree that we forget about the other kind of risk. Type I error is the risk of "crying wolf," just like the old fable of "The Boy Who Cried Wolf." When we commit Type I error, we claim that the effect of a variable is important, but it turns out that it isn't a finding that will generalize to the population.

Type II error is the risk that there is an amazing finding awaiting us in the population, but our analysis of the sample data doesn't reveal it. That's pretty bad, too — claiming that there is no effect when there really is one. The Sample Power module allows us to accurately calculate that risk, and it may prompt us either to collect more data to avoid the risk, or maybe, just maybe, we figure out that we can get by with a little less data and we can save our organization money during the data collection phase.

If you do survey research and have to go out into the world to collect your data, this module is one that you'll want to consider. Even if you get your data through other means — the day-to-day running of the business, for instance — look up Jacob Cohen on the web and seek out his writing. One of our favorites is *Things I Have Learned (So Far)*. Don't obsess over 0.05 or forget about Type II error.

The Visualization Designer Module

The Visualization Designer module doesn't get as much attention as it deserves. Even veteran SPSS users don't seem to know that much about it. Graphboard Template Chooser is one of the graphing methods in SPSS, and this module is actually a sibling product to Graphboard in a sense.

If you want to create really fancy graphs in SPSS, you have two choices: Learn how to program Graphics Production Language (GPL) or use the Visualization Designer module. GPL isn't really that bad, but for some folks, writing code just isn't their thing. The Visualization Designer module allows you to create all kinds of graphics that aren't possible otherwise, and when you're done, you can add new "templates" to your copy of SPSS and to that of your colleagues, too. When you're done, the new templates will show up as new chart types in the Graphboard Template Chooser.

Chapter 23

Ten (Or So) Useful SPSS Online Resources

In This Chapter

▶ Connecting with other SPSS users on the Internet

▶ Finding the information you need online

*I*BM SPSS Statistics users are all over the world. The Internet is a powerful medium through which you can join the SPSS community, and this chapter points you in the right direction.

The Statistics & Consultants Group on LinkedIn

The Statistics & Consultants Group on LinkedIn is one of the largest groups of its kind. At the time of this writing, there are more than 40,000 members. One of your authors, Keith McCormick, is one of the group's managers. Almost every day, a new discussion is initiated on some aspect of statistics, statistical software, or professional development. Joining the group is completely free and well worth your time. Before you know it, you'll be not only reading the discussions, but also initiating them and eventually answering other people's questions.

To join, visit the group's profile page on LinkedIn: www.linkedin.com/groups/Statistics-Analytics-Consultants-Group-1592517/about. Note that the group is private so you need to submit a request to join, but requesting membership is easy — just have a fairly complete LinkedIn profile. There are no special statistical qualifications necessary.

Here are some pointers for getting started with the group:

- ✔ **Start by reading.** You don't have to post or comment to get a lot of value out of a professional group like this.

- ✔ **When you're ready to contribute, be a good group citizen.** Try to comment only when you think you can add value. Provide an appropriate amount of context when starting a discussion. Make it interesting, inviting lots of participation. Questions like "How do you do Factor Analysis?" are too vague. Members are there to learn (and teach), and they sincerely want to help.

- ✔ **If your agenda goes beyond just learning, and you want to hire (or be hired) or promote a product or service, check out the special areas for jobs and promotions.** When appropriate, participate in those areas. Be respectful of reasonable boundaries between the different areas by placing promotional content in "promotions" and not in the discussion area.

- ✔ **Join the SPSS subgroup as well.** There is a smaller group dedicated to just SPSS Statistics. There is tons of SPSS activity in the big group, but the subgroup is a cozier setting for discussions specific to SPSS software.

SPSSX-L

SPSSX-L (which you can find at `https://listserv.uga.edu/cgi-bin/wa?A0=SPSSX-L`) has been around for years and years. In fact, it predates most of the other resources you'll find on the Internet. SPSSX-L is a listserv, which is an email-based posting system. You send an email to join, and the posts come back to you in the form of email.

You may find the listserv format surprising if you're young enough that you've always had the Internet. Even if the idea of a listserv is a bit quaint, you don't want to miss out on the wisdom that's available through this group. Some of the most knowledgeable and veteran SPSS users out there are active in SPSSX-L, and they sincerely want to help other users.

To ask a question, simply send an email (instructions are given on the home page) and the system will forward your email to all the members. There are instructions on how to cancel if you find that the flow of email is more than you like. You might want to set up an email filter where all the messages from this listserv go to a special folder in your email system, so you can read them when you have the time.

IBM SPSS Statistics Certification

If you're in a corporate setting or looking to get a corporate job using SPSS, it may help to get certified in SPSS. Having this certification listed on your LinkedIn profile or résumé may help when you're ready to transition to another role or organization. Currently, you can find details about the IBM SPSS Statistics Level 1 certification exam at `www-03.ibm.com/certify/tests/objC2020-011.shtml`.

If you're in a university setting, than this test probably wouldn't be as useful as taking (and doing well) in a university class that uses SPSS. At the time of this writing, there are five approved Global Training Providers (GTPs) in the IBM training economy:

✔ Arrow ECS Education

✔ Avnet Academy

✔ Global Knowledge

✔ Ingram Micro Training

✔ LearnQuest

Online Videos

You can find both free and for-fee videos online. You can find full video case studies from IBM at `www-01.ibm.com/software/analytics/spss/downloads/demos.html`. Check out "Online Demo: Business Analytics Statistics Overview Demonstration" and "Online Demo: IBM SPSS Custom Tables in Action."

If you want to go the for-fee route, video courses are a great option. Here are some you may want to check out:

✔ **Lynda.com (`www.lynda.com`):** Lynda.com is a popular website for video training. It offers training for all kinds of popular software products, and has a subscription-style pricing structure. They offer one SPSS Statistics course, "SPSS Statistics Essential Training," which is popular and well done.

✔ **Udemy (`www.udemy.com`):** Udemy offers several SPSS classes for a variety of topics and at varying lengths and prices. "IBM SPSS Statistics: Getting Started" was created by the authors of this book.

Twitter

TIP

Twitter might seem like a strange suggestion at first. After all, what can you learn about SPSS in just 140 characters? Of course, the tweet itself won't help, but what you'll find on Twitter are thought leaders in the SPSS community and the latest and greatest information on SPSS.

Here are some recommendations on who to follow:

- ✔ **IBM Training (@IBMTraining):** Official news on IBM training including SPSS Training.

- ✔ **IBM SPSS Software (@IBMSPSS):** Official IBM SPSS tweets.

- ✔ **developerWorks (@developerworks):** The latest news on SPSS programming.

- ✔ **IBM Insight (@IBMInsight):** Tweets about the big annual IBM conference in Las Vegas that features a lot of info about SPSS.

- ✔ **Keith McCormick (@KMcCormickBlog):** Keith is one of the authors of this book. Check out some of the folks that Keith follows to get more suggestions.

- ✔ **Jon Peck (@jkpeck):** An employee of SPSS, Inc. (now IBM) for most of SPSS's history. His tweets are a great way to find out about the latest new programming features.

- ✔ **Bob Muenchen (@BobMuenchen):** Author and speaker who specializes in teaching SPSS users (and SAS users) about R.

- ✔ **Armand Ruiz (@armand_ruiz):** A young IBMer who is an emerging IBM leader in SPSS programming.

Here are some Twitter accounts that are great for data visualization and analytics in general:

- ✔ **The American Statistical Association (@AmstatNews):** The American Statistical Association is the world's largest community of statisticians.

- ✔ **SignificanceAp Magazine (@signmagazine):** A joint statistics magazine from @RoyalStatSoc and @AmstatNews. Great source for interesting articles both in an out of the magazine.

- ✔ **Simply Statistics (@simplystats):** Simply Statistics blog by Jeff Leek, Roger Peng, and Rafael Irizarry.

- ✔ **Nathan Yau (@flowingdata):** Very influentially author and blogger specializing in Data Visualization.

- ✔ **Edward Tufte (@EdwardTufte):** Famous for his data visualization books and harsh critique of PowerPoint presentations.

- ✔ **Gregory Piatetsky (@kdnuggets):** Gregory I. Piatetsky is a Data Scientist, co-founder of KDD conferences and ACM SIGKDD association for Knowledge Discovery and Data Mining, and President of KDnuggets, a leading site on Business Analytics, Data Mining, and Data Science.

- ✔ **Meta Brown (@metabrown312):** Author of *Data Mining For Dummies* and a thought leader in predictive analytics.

- ✔ **Hans Rosling (@HansRosling):** Brilliant lecturer famous for his TED talks.

- ✔ **Nancy Duarte (@nancyduarte):** Author of *Slideology*. Owner of a successful company that polishes corporate presentations. She became famous when she helped Al Gore with his slide presentations.

- ✔ **Dean Abbott (@DeanAbb):** Well-known data miner who speaks at the Predictive Analytics World conferences.

- ✔ **Andrew Ng (@AndrewYNg):** Chief Scientist of Baidu; chairman and co-founder of Coursera; Stanford CS faculty.

- ✔ **Gil Press (@GilPress):** Tech journalist. Everything he writes is worth a quick read.

And here are some that are just fun to try:

- ✔ *The New Yorker* **(@NewYorker):** If you're going to try Twitter, you can't go wrong with tweets from the world's best magazine.

- ✔ **Maria Popova (@brainpicker):** Random thoughts on all kinds of interesting things. More arts and literature than stats, but fun nonetheless.

- ✔ **TED Talks (@TEDTalks):** Tweets from the same folks who bring you those great 18-minute long video talks.

Blogs

Blogs can be a great way to get current advice on SPSS. Books may be updated every time a new version of SPSS is released, but between versions new tips and tricks materialize and that's where blogs can be a great resource.

Not all blogs are created equal, but here are some we recommend:

- ✔ **AnalyticsZone (www.analyticszone.com):** AnalyticsZone is an IBM blog that has a lot of good current material. They frequently announce new plug-in code that you can borrow to improve the functionality of your copy of SPSS.

✔ **Keith McCormick** (`www.keithmccormick.com`): This is the blog of one of the authors of this book, and although he can't promise a post every week, he has posted lots of useful advice over the years. Some of it is at an intermediate level that will be useful after you've read this book.

✔ **Raynald's SPSS Tools** (`www.spsstools.net`): This SPSS-related blog that has been popular for decades.

Most blogs have a *blog roll* (a list of other blogs that particular blogger recommends), so when you start reading one blog you can find others that are similar. Twitter can be a great way to find more blogs, too (see the preceding section). Most bloggers announce their new posts on Twitter.

Online Courses with Live Instruction

There are plenty of instructors out there waiting to teach you a live class. Many of these classes are quite reasonably priced. Both Jesus Salcedo and Keith McCormick (two of your authors) regularly teach online, sometimes to audiences halfway around the world. Jesus's training schedule can be found at `https://learn.quebit.com/category/spss-statistics-courses`. You can find out more about Keith's schedule at (`keithmccormick.com/SPSSTraining`).

They both teach some theory classes, but they can always be counted on to offer introductory and advanced point-and-click classes as well, focusing on software operations.

There is a whole economy of SPSS software instruction out there. Make sure to find out who's doing the actual teaching, and don't be shy about emailing them or chatting with them before you choose a class. You can always email Keith (`keithmc123@gmail.com`) and Jesus (`jesussalcedo@yahoo.com`) for advice. They know most members of the SPSS community.

One interesting option for learning SPSS Statistics is a brown-bag lunch format that is offered by The Analysis Factor (`www.theanalysisfactor.com`). They also offer more extensive seminars on a variety of topics. They cover advanced topics in nonthreatening, shorter formats, too. The training isn't limited to SPSS Statistics, but SPSS content is common.

A very extensive list of course options is awaiting you at Statistics.com (`www.statistics.com`). These courses are like university short courses with homework and the whole nine yards. The classes are typically asynchronous so you might see recordings of lectures, but you'll have access to the instructor during the multiple weeks of the course. A long list of statistics professors and textbook authors are among their ranks. Statistics.com offers serious, in-depth classes, which may be just what you're looking for. We can't guarantee that SPSS Statistics will be the software tool of choice, so if that's important to you, check before you enroll. If the class sounds perfect, but it's taught using another software tool, don't rule it out.

Finally, Coursera (`www.coursera.org`) has become especially popular in recent years. You're much more likely to find a course about statistics than you are one specifically about SPSS Statistics software operation, but more classes are added all the time, so check it out!

Tutorials

Finding free SPSS content on the Internet isn't difficult. The challenge is finding *good* free content. The folks at UCLA have maintained a great website with tutorials for years; you can find it at `www.ats.ucla.edu/stat/spss`. We highly recommend UCLA's tutorial.

Here are two more that are worth checking out:

- **Amherst College** (`www.amherst.edu/academiclife/departments/ psychology/resources/SPSS`): Amherst's tutorials include a lesson on APA format.
- **London School of Economics** (`www.lse.ac.uk/methodology/ tutorials/SPSS/home.aspx`): The tutorials from the London School of Economics make it easy to join in with the examples by providing the practice data for easy download.

SPSS Programming and Data Management: A Guide for SPSS and SAS Users

SPSS Programming and Data Management: A Guide for SPSS and SAS Users, by Raynaud Levesque and SPSS, Inc., is a great book. As of this writing, a new updated version for SPSS Statistics 23 was just released. (The best place for information on python and SPSS is `www.ibm.com/developerworks`.) Levesque's book is best for giving you solid intermediate-level instruction.

You may find that the SPSS Statistics Syntax chapters in Levesque's book are advanced, but once you get to an intermediate level in your knowledge of SPSS Statistics Syntax seek out this free book. You can download this book for free at www.spsstools.net/SPSS_Programming/SPSS%20Programming%20and%20Data%20Management%203rd%20Edition.pdf. Or just search the web for the book's title and you'll find it.

Chapter 24

Ten Professional Development Projects for SPSS Users

In This Chapter

▶ Challenging yourself with more advanced SPSS projects

▶ Taking your SPSS knowhow to the next level

*W*hen you're ready for intermediate-level material, this chapter is for you. Here, we take you beyond what this one book can do, and introduce you to a whole world of SPSS knowledge out there waiting for you.

For some of these topics, *SPSS Statistics for Data Analysis and Visualization*, by Keith McCormick and Jesus Salcedo (Wiley), can be a big help.

The Case Studies

The case studies in the Help menu are an absolutely wonderful resource, and many new users simply don't find them. They're chapter-length, step-by-step, detailed walk-throughs of the techniques, accompanied with practice data. In terms of free resources, you simply can't get any better than this. The only limitation, perhaps, is that for the more sophisticated techniques, you encounter a lot of jargon (so you may find yourself searching for more information online).

If you happen to have a good grounding in statistical jargon already, jump right in! If you're just starting out, start with the case studies that support the techniques in Part V of this book. As for the jargon, this book's Glossary will get you started.

Syntax

Syntax is a big topic, and for many people (okay, most!), it's a little dry. But syntax is very powerful, and you won't learn it by osmosis. SPSS doesn't force you to learn syntax — you have to force yourself. The medicinal taste won't go away right away, but soon you'll be saving time, and that will motivate you to keep going! At first, it may seem like it's taking *more* time, but your persistence will pay off.

Here are a few things you can do to get started:

✔ Go to any procedure under the Data, Transform, or Analyze menus. Complete the dialog box as if you were going to run the technique. Before you click OK, click the Paste button. The Syntax appears in the Syntax Editor. Now you can look up the command in the Syntax Reference Guide (located within the Help menu) to learn more about it.

✔ Take a simple formula like body mass index (BMI), gross domestic product (GDP), or something from your field, and try doing the calculation in the Syntax window. Eventually, you'll find that many calculations are faster this way, especially if you're doing lots of similar calculations, which is very common in the SPSS world.

✔ Read the opening chapter of the Syntax Reference Guide (located within the Help menu). There are a couple of other chapters worth taking a peek at, but you can mostly use it as a dictionary. Don't try to read the whole thing — it's a door stopper.

✔ Create a simple project: open a file, add labels, add a variable or two, produce a table, produce a graph, and each time before you click OK, click the Paste button, which produces the Syntax in the Syntax Editor. Now you have a copy of the Syntax and you can run it, but more important, you can save it so you can redo all these techniques at a later point (with new data). Force yourself. It may take a few hours or even a whole day. You might hate us for suggesting it, but you'll be glad you followed through. Soon, you'll be working at three times your old speed.

IBM SPSS Tables

For years and years, the Table module has been the most popular module in SPSS. And you can easily see why! The Table module is easy and powerful. Plus, marketing and survey researchers just love it, and they've made it popular. You don't have to work in either of these areas to benefit from the Table module, though.

TIP

Get a free trial copy of the complete SPSS Statistics with all the modules, and force yourself to spend a solid day using it. See if there is any aspect of reporting that you're already doing that you could do faster with the Table module. Force yourself to reproduce a recent report, and see how much time you might save.

The following two figures were in previous chapters. In Figure 24-1, you see a simply Frequencies showing two variables. Note that the categories for both variables are the same. In Figure 24-2, you see the same data, but here the table was created using the SPSS Table module: this is a much better table.

Frequency Table

Speaker_Discount

		Frequency	Percent	Valid Percent	Cumulative Percent
Valid	Discount	1645	49.3	49.3	49.3
	Regular	1693	50.7	50.7	100.0
	Total	3338	100.0	100.0	

Stereo_Discount

		Frequency	Percent	Valid Percent	Cumulative Percent
Valid	Discount	1022	30.6	30.6	30.6
	Regular	2316	69.4	69.4	100.0
	Total	3338	100.0	100.0	

Figure 24-1: Frequencies table of the discount variables.

Figure 24-2: Custom table of the discount variables.

Custom Tables

	Discount	Regular	Total N
Speaker_Discount	49.3%	50.7%	3338
Stereo_Discount	30.6%	69.4%	3338

If you're producing the table for yourself, it doesn't matter. But if you're putting the table in a report that will be sent to others, you really need the SPSS Table module. By the way, with practice it takes only a few seconds to make the custom version, and you can use Syntax to make it even better!

Data Visualization

This topic has inspired a cottage industry of advice. Unfortunately (or maybe fortunately), they aren't all SPSS point-and-click books. Why fortunately? Well, it may be a good idea to broaden your horizons and think about good design, and not worry, at first, about how you're going to make it. Have a little faith that SPSS can do it. Here are some resources to get you started:

- *The Visual Display of Quantitative Information,* by Edward R. Tufte (Graphics Press), is a classic.

- *Visualize This: The Flowing Data Guide to Design, Visualization, and Statistics,* by Nathan Yau (Wiley), is great. FlowingData is the name of Yau's blog, which is also excellent.

- *The Wall Street Journal Guide to Information Graphics: The Dos and Don'ts of Presenting Data, Facts, and Figures,* by Dona M. Wong (W. W. Norton & Company), is a great book written by an admirer of Tufte. You may even want to read this book before Tufte's because it's shorter and can be applied to your work more easily.

- Hans Rosling is a master of story telling with statistics. His presentations can be found at www.ted.com/speakers/hans_rosling. Note Rosling's use of the bubble chart style of graphic (see Chapter 12), which he has helped popularize. Impressive charts resembling what Rosling shows (without the animation) can be done with Graphics Production Language in SPSS.

Better Presentations

You'll likely have to present your analysis in a slide presentation someday. Most slide presentations are terrible, and you don't want to add to the world's inventory of bad presentations. You simply can't be an exceptional SPSS Statistics user if no one understands what you're trying to say about your results.

Check out the following books, and then spend an afternoon watching TED videos with some ice cream or popcorn or beer:

- *slide:ology: The Art and Science of Creating Great Presentations,* by Nancy Duarte (O'Reilly Media), in many ways started the wave of writing on this subject. *Resonate,* also by Nancy Duarte (Wiley), is an equally good companion piece.

✔ *Presentation Zen: Simple Ideas on Presentation Design and Delivery,* 2nd Edition, by Garr Reynolds (New Riders), is full of great advice.

✔ *Talk Like TED: The 9 Public-Speaking Secrets of the World's Top Minds,* by Carmine Gallo (St. Martin's Press), has reduced the secrets of the 18-minute TED-style presentation to nine key points.

✔ *The Cognitive Style of PowerPoint: Pitching Out Corrupts Within,* 2nd Edition, by Edward R. Tufte (Graphics Press), is a deservedly famous 32-page rant against bad PowerPoint.

R

That isn't an abbreviation. It's the whole name of a programming language. If you're a statistics major or minor, you've almost certainly heard of it. R is incredibly popular. You may have even had friends compare R to SPSS and talk about how it's powerful and free. R *is* powerful and free, but there is no need to make it compete with SPSS. You can use R right in the Syntax window of SPSS, giving you the best of both worlds.

R has a much steeper learning curve than SPSS does. Many books on R focus on the basics, but as an SPSS user, you don't need all that. The basics are more easily done in SPSS. The power of R in the Syntax window is that you'll never be without a feature that you need in SPSS. If something brand new comes out, and it often appears in R first, you can access it through SPSS while still taking advantage of the very easy point-and-click interface of SPSS. Make no mistake: R is programming.

Graphics Production Language

Graphics Production Language (GPL) is programming, and we know that for some people, that's a strike against it. GPL is powerful, though. It will transform your notion of what is possible with SPSS graphics. GPL is probably the only way to be able to follow the advice of folks like Nathan Yau and Edward Tufte in SPSS. Sure, you can take their advice and do all kinds of clever things, and make some major improvements to your graphics, but if you really want to get serious about it, you'll have to learn how to code. That's true of everyone who wants to do serious graphics.

Output Management System

Output Management System (OMS) changes everything. When you grow to be a sophisticated user of SPSS, there is simply no reason to be manually cutting and pasting all your results, one at a time. That's a frequent mistake new users make as they learn more and more SPSS. Cutting and pasting makes a lot of sense when you have one table or one graph, but you don't want to use this approach for 30 tables or 81 graphs.

The whole idea behind OMS is that you can automatically route results to just about any format in one step just by telling SPSS what you want routed and where you want it to go. Next thing you know, your Regression Q-Q Plots will all go to PowerPoint. *Voilà!* There's more to OMS, of course, but not so much that you can't figure it out in a few hours or less. If you produce a lot of output, learn this approach as soon as possible. It may save you many days each year.

Python Programs

Years ago, and we mean many years ago, there was a feature in SPSS called Macros. Some folks still like Macros, just like some folks still like making ice cream by manually turning a crank. Python is the way to go, if you want to be a true power user when it comes to SPSS programming. Have no fear. If you enjoy programming, and you're feeling brave, a free book called *SPSS Programming for SPSS and SAS Users* can help. As of this writing, the most recent update is for version 20, but it will give you what you need.

In the SPSS Help, you'll also find Python Integration Package for IBM SPSS Statistics and Introduction to Python Programs, both of which are great resources.

Python Scripting

Python scripting is quite different from Python programming in SPSS Statistics. Scripting has its own Help section, Python Scripting Guide for IBM SPSS Statistics. The main difference is that Python programs allow you to do more powerful things with syntax. Python scripts help you manipulate the interface — the output window, results in the output window, the graphical user interface, and so on. Scripts are very powerful, but if you aren't a programmer, start with Python programs first.

Glossary

add-on: A utility that can be added to SPSS. Also called a module.

adjusted R Square: Represents a technical improvement over R Square in that it explicitly adjusts for the number of predictor variables relative to the sample size.

alternative hypothesis: A hypothesis that states that an effect is present. *See also* null hypothesis.

ascending: A sorting order. The cases are ordered so the values range from small to large. *See also* descending.

B coefficients: In linear regression, they show how a one-unit change in an independent variable impacts the dependent variable.

base: The main system of SPSS. Modules can be added to expand SPSS, but the base system is always present.

bell curve: *See* normal distribution.

betas: In linear regression, they are standardized regression coefficients and are used to judge the relative importance of each of the independent variables.

binning: The process of organizing the values of a variable into groups. Each group is a defined as a specific range of values.

bivariate: An analysis using two variables.

case: All the values in a single row. A case is sometimes called a single record.

case summary: A simple table that directly summarizes values of the cases.

categorical variable: A type of variable that has values that are qualitatively different, such as gender, eye color, and region of the country. *See also* ordinal variable *and* nominal variable.

chart: *See* graph.

confidence interval: A range of values above and below the mean into which a specified percentage of the values appears. For example, if gravel trucks for a company deliver an average of 190 loads per month, but 95% of the trucks deliver between 186 and 194 loads, the 95% confidence interval ranges from a low of 186 to a high of 194.

continuous variable: See *scale variable.*

control variables: Additional variables for which you remove their impact in an analysis.

correlation: The degree of similarity or difference between two variables.

crosstabulation: Used to study the relationship between two or more categorical variables.

cutpoint: A number used as a divider to split values into groups, as in binning.

dataset: The data displayed in the Data Editor window, whether loaded from a file, entered from the keyboard, or both.

delimiter: A character used to indicate the beginning of, ending of, or separation between individual values in a series of strings of characters. For example, the string of characters 59,21,34 is a series of comma-delimited numbers.

dependent variable: A variable that has its value derived from one or more other variables. *See also* independent variable.

descending: A sorting order that arranges values from large to small. *See also* ascending.

descriptives procedure: Provides a succinct summary of various statistics and the number of cases with valid values for each variable included in the table.

deviation: The amount by which a measurement differs from some fixed value.

dichotomy: A variable with only two possible values, such as yes/no, true/false, or like/dislike. It is a specific type of categorical variable.

false negative: Falsely concluding that there is not a significant difference when in reality there is a difference. Also known as a Type I error.

false positive: Falsely concluding that there is a significant difference when in reality there is no difference. Also known as a Type II error.

field: In the SPSS documentation, *field* is used as a synonym for *variable*.

frequency distribution: The collection of values that a variable takes in a sample.

frequency table: Provides a summary showing the number and percentage of cases falling into each category of a variable.

goodness of fit: The extent to which observed values approximate values from a theoretical distribution.

graph: A nonnumeric display of values. The terms *graph* and *chart* are used in SPSS internal documentation almost interchangeably.

graphical user interface (GUI): Control of an application with windows and a mouse. All versions of SPSS operate this way.

GUI: *See* graphical user interface.

histogram: A graphical display of a distribution in which the extent of each bar represents the magnitude (as in a bar chart) and the width of each bar represents the magnitude of the bin. The area of each bar thus represents the frequency.

hypothesis testing: Making an inference about a population from a sample.

independence: The degree to which two or more variables have no effect on one another.

Independent-Sample T Test: A test that determines whether the means for two groups differ on a continuous dependent variable.

independent variable: A variable whose values are used to predict the values of a dependent variable. *See also* dependent variable.

interval variable: A variable where a one-unit change in numeric value represents the same change in quantity regardless of where it occurs on the scale.

kurtosis: A measure of how peaked a distribution is. A positive number indicates there is more of a peak than standard; a negative number indicates a flatter line.

levels of measurement: Refers to the coding scheme or the meaning of the numbers associated with each variable.

Levene test: A test that determines whether the variance of two groups is significantly different or the same.

linear: A straight line.

linear regression: Tries to predict the values of one variable based on another.

maximum: The highest value for a variable.

mean: The mathematical average of all the values in the distribution (that is, the sum of the values of all cases divided by the total number of cases).

means procedure: Calculates subgroup means and related statistics for dependent variables within categories of one or more independent variables.

median: The midpoint of a distribution; it's the 50th percentile.

minimum: The lowest value for a variable.

missing data: If you declare a value for a variable as representing the fact that no value is present, the missing value will not be included in calculations.

mode: The category or value that contains the most cases.

module: *See* add-on.

multiple-response set: A special variable that has its content generated from the content of two or more other variables. In SPSS, it doesn't appear on the Data View tab of the Data Editor window, but it does appear when you select variable names for other menus.

multivariate: An analysis using more than two variables.

nominal variable: Values that represent categories. There is no inherent order to the categories. For example, yes, no, and undecided could be represented by 2, 1, and 0. *See also* scale variable, ordinal variable, *and* categorical variable.

nonlinear: Not in a straight line.

normal distribution: A distribution that is continuous and symmetric. It is used primarily because many quantitative measurements appear to approximate this distribution. Also called the bell curve.

normality: The degree to which the values match the normal distribution.

null hypothesis: A hypothesis that states that no effect is present.

OLAP cubes: *See* online analytical processing cubes.

online analytical processing (OLAP) cubes: A multilevel table containing totals, means, or some other statistics in which each level of the table contains the values relating to one value of a categorical variable.

OMS: *See* output management system.

One-Sample T-Test procedure: Tests whether the mean of a single variable differs from a specified value.

One-Way ANOVA: Tests whether the means for two or more groups differ on a continuous dependent variable.

ordinal variable: Variables where there is a meaningful order or rank to the categories. However, with ordinal data there is not a measurable distance between categories. The ordinal forms of 1, 2, and 3 are first, second, and third. *See also* scale variable, nominal variable, *and* categorical variable.

outliers: The extreme values of a variable. Generally, these are cases that are more than three standard deviations away from the mean.

output management system (OMS): The ability in SPSS to output to different file formats.

Paired-Samples T Test: Tests whether means differ from each other under two conditions.

paneling: Adding another dimension of data to a graphic display causing the layout to be replicated a number of times to accommodate the values of the data along the new dimension.

Pearson correlation: Represents the degree of linear relationship between two continuous variables.

Pearson Chi-Square: Determines whether there is a relationship between two categorical variables.

pivot table: Tables that users can edit. The tables in the SPSS Statistics Viewer window are pivot tables.

predicted variable: *See* dependent variable.

predictor: A variable, or collection of variables, that predict the values of a dependent variable.

quartile: Specific values that divide all the values into four groups, with an equal number of values in each group. The groups are generally called the first, second, third, and fourth quartiles.

R: The correlation between the dependent measure and the combination of the independent variable(s). Therefore, the closer R is to 1, the better the fit.

R Square: The correlation coefficient squared. It can be interpreted as the proportion of variance of the dependent measure that can be predicted from the combination of independent variable(s).

range: The difference between the maximum and minimum values.

ratio variable: Variable that has all the properties of interval variables with the addition of a true zero point, representing the complete absence of the property being measured.

record: All the values of a single row. It is a single case or row.

row: A single row in the Data View tab of the Data Editor window. It is a single case.

scale variable: A type of number that uses a standard by which something is measured, such as inches, pounds, dollars, or hours. Another name for scale is *continuous. See also* ordinal variable, nominal variable, *and* categorical variable.

script: A program written in either the BASIC or Python programming language. These languages are different from Syntax.

standard deviation: The variance measure is expressed in the units of the variable squared. This can cause difficulty in interpretation, so more often, the standard deviation is used. The standard deviation is the square root of the variance, which restores the value of variability to the units of measurement of the original variable.

standard error of the estimate: Provides an estimate (in the scale of the dependent variable) of how much variation remains to be accounted for after the prediction equation has been fit to the data.

Summary Independent-Sample T Test: Uses summary statistics to test whether the means for two groups differ on a continuous dependent variable.

Syntax: The name of the programming language fundamental to SPSS. All actions performed by SPSS are in response to the internal interpretation of Syntax commands.

univariate: A statistic derived from the values of one variable. Examples are mean, standard deviation, and sum.

variance: Provides information about the amount of spread around the mean value. It is an overall measure of how clustered data values are around the mean. The variance is calculated by summing the square of the difference between each value and the mean and dividing this quantity by the number of cases minus one. In general terms, the larger the variance, the more spread there is in the data; the smaller the variance, the more the data values are clustered around the mean. The variance is the square of the standard deviation.

Index

• P •

paneling, 166, 351. *See also* graphs
PASW (Predictive Analytics Software), 10
.pdf files, 79
PDF files
  advantages, 93
  export options, 93
  HTML versus, 92
Pearson chi-square test. *See* chi-square test
Pearson correlation coefficient
  assumptions, 255
  definition, 351
  linear relationships, 252, 350
  variable types compared, 249
perceptual maps, 326
period (.), 312
pie charts
  creating, 194
  label display, 26–27
  Summaries of Groups of Cases, 38–39
  variables selection, 39
pivot tables
  Excel files, 88
  HTML web page files, 86
  layers
    Chi-Square Tests table, 233
    control variables, 230–232
    overview, 86, 302
    PDF file inclusion, 93
    pivoting trays, 301–302
    PowerPoint options, 91
    stacked area charts, 185
    Word output choices, 89
  overview, 293, 351
  PDF files, 93
  pivoting trays
    crosstabs selection, 299–300
    file selection, 299
    layers, 301–302
    overview, 299
    setting up, 300–301
    tray positioning, 301
  PowerPoint slides, 91–92
  settings, 275–276
  Style Output feature
    Crosstabs dialog box use, 297
    file selection, 297

main dialog box use, 297–298
    overview, 297
    Table Style dialog box use, 298–299
  TableLooks feature
    Cell Display dialog box use, 294
    context menus, 295
    Crosstabs dialog box selections, 293–294
    file selection, 293
    TableLooks dialog box use, 295–296
Pivot Tables tab, Options window, 275–276
.png files, 95
population pyramids
  categorical variable use, 184
  creating, 183–184
  overview, 183
.por files, 74
PowerPoint slides, 91–92
.ppt files, 80
Predictive Analytics Software (PASW), 10
Presentation Zen: Simple Ideas on Presentation Design and Delivery (Reynolds), 345
presentations, 344–345
probability testing. *See also* analysis; inferential testing; t tests
  alpha levels, 219
  null versus alternative hypotheses, 219, 236, 350
  overview, 218, 349
  statistical outcomes
    false positives/negatives, 237, 348
    significant versus not significant, 236
profiling, 330
Python
  Essentials for Python installation, 18–19
  programs versus scripts, 11, 352
  SPSS Help, 346
  SPSS Programming for SPSS and SAS Users, 346

• Q •

QUEST algorithm, 327

About the Authors

Keith McCormick: Keith moved to Raleigh-Durham, North Carolina, in the late '90s planning on graduate school, but a "part time" SPSS training job turned out to be a career. Keith has been all over the world training and consulting in all things SPSS, statistics, and data mining. Several thousand students later, hopefully, he has learned a thing or two about how to get folks started in SPSS. These days, he splits his time between doing SPSS Statistics and SPSS Modeler work, teaching others how to do it, and writing about how to do it. Periodically, he puts the latest news on his blog at KeithMcCormick.com. In between assignments, Keith likes to travel to out-of-the-way places, try new food, hike around, and find cool souvenirs to bring home.

Jesus Salcedo: Jesus is QueBIT's Director of Advanced Analytics Training. Previously, he worked for IBM SPSS as the SPSS Curriculum Team Lead and as a Senior Education Specialist. He has been using SPSS Statistics for 20 years. He has many years of teaching and consulting experience and has taught statistics and data mining to thousands of students using both SPSS Statistics and SPSS Modeler. He earned a PhD in psychometrics from Fordham University. In his free time, Jesus enjoys playing baseball, hiking, and traveling.

Aaron Poh: Aaron is a management and technology consultant working with federal and commercial clients. His experience ranges from IT strategy and enterprise architecture development to organizational design and business process reengineering. He holds a master's degree in engineering management from Duke University and a BS in biomedical engineering and applied mathematics from Johns Hopkins University. An active participant in various leadership roles, he strives to impact the organizations and people around him in a positive way. Aaron is a self-motivator and enjoys developing creative solutions to further organizational success and promote positive cultures and campaigns. In his free time, he loves to travel and experience new cultures and foods.

Dedication

To all the "legacy SPSS" folks we've known, and worked with, over the years, some of whom are at IBM now, and some of whom have moved onto other things.

Authors' Acknowledgments

We owe our thanks to Meta Brown for introducing us to the team at John Wiley & Sons. Amy Fandrei stewarded the book for the third time and helped us overcome obstacles along the way. Thank you, Amy. We would also like to thank our editor, Elizabeth Kuball; our technical editor, Shannon Johnston; and the rest of the *For Dummies* team.

We would like to thank all our friends and colleagues at IBM SPSS.

Keith and Jesus would like to thank Aaron for his contributions. Without him, we couldn't possibly have been ready in time for the Version 23 release. We're proud that this book is absolutely up to date with the latest and greatest, and Aaron helped us get it done.

Finally, we owe a debt to Arthur Griffith, who wrote the first and second editions. Although we never had an opportunity to meet Arthur, his presence is felt in these pages. Just when we might risk taking ourselves too seriously, we stumble upon a phrase of Arthur's featuring "diddlysquat" or a software illustration using jelly beans. His wonderful sense of humor has been retained whenever possible in the revision, since we couldn't possible replicate it.

Publisher's Acknowledgments

Acquisitions Editor: Amy Fandrei

Project Editor: Elizabeth Kuball

Copy Editor: Elizabeth Kuball

Technical Editor: Shannon Johnston

Production Editor: Vinitha Vikraman

Cover Image: © iStock.com/3alexd